Joseph F Green

Ocean Birds

Joseph F Green

Ocean Birds

ISBN/EAN: 9783337065539

Printed in Europe, USA, Canada, Australia, Japan

Cover: Foto ©berggeist007 / pixelio.de

More available books at **www.hansebooks.com**

OCEAN BIRDS.

BY

J. F. GREEN.

WITH

A PREFACE BY A. G. GUILLEMARD.

AND

A TREATISE ON SKINNING BIRDS BY F. H. H. GUILLEMARD, M.D.

The Illustrations by

FRANCES E. GREEN.

LONDON:

R. H. PORTER, 6, TENTERDEN STREET, W.

1887.

CONTENTS.

LIST OF PLATES.

(Drawn by Frances E. Green.)

PREFACE.

THE FOLLOWING NOTES ON OCEAN BIRDS, their appearance and habits, have been compiled with the special object of interesting, and at the same time, to some small extent, instructing a special class of readers—those who are led by business, or pleasure, or the pursuit of health, to take a long sea voyage. And it may be stated at the outset that the birds depicted and described in the following pages are those usually seen in the course of a voyage from England to Australia or New Zealand round the Cape of Good Hope, and home round Cape Horn, thus completing a tour round the world. Those of us alone who have made a voyage of this kind can appreciate fully the interest which attaches to Ocean Bird-life.

Every passenger who embarks on board a ship bound for Australasia, is well aware that the voyage in prospect must, from more than one point of view, be monotonous. For a period of some three months he will be cribbed, cabined, and confined within the narrow limits of the ship's decks, and restricted as to society to a small circle of fellow-passengers, amongst whom it must be hoped that he will find a few with tastes and trains of thought congenial to his own. In point of exercise he will be limited to his daily constitutionals on the poop or in the waist of the ship, and to occasional climbs aloft, upon which latter he will venture with no slight amount of trepidation. Amusements will be open to him in the shape of the ordinary games, such as cock-fighting, boxing, and small cricket, that can be carried on on deck, and chess tournaments, theatricals, and concerts will occupy pleasantly many a lazy hour. If wise, he will not fail to take with him a goodly store of standard books, for one always finds on board ship more time for reading than falls to one's lot when ashore. Stirring events

b

may possibly be in store for him, but the majority of passengers will be found to confess that they are not able to recall to memory more than a dozen occurrences a few years after their voyage has terminated.

A gale in the dreaded Bay of Biscay, a glimpse of Madeira's mountains rising abruptly from the blue plain of the sea, the first gorgeous sunset in the Tropics, the first "school" of Flying-fish scudding away from the ship's side, a "clock calm" on the Line, the first mighty Albatross soaring majestically over the poop, a fleeting vision of the glorious Peak of Tristan da Cunha towering up into the cloud-wreaths, the first sight of the Southern Cross, the first Whale, the first look at Australia,—events and scenes such as these will remain as the most noteworthy memories of a voyage to the sunny South. It will readily be gathered that these events and scenes, being mostly of transient interest, do not go far towards varying the monotony of sea and sky during a period of three months. And mainly for this reason so soon as the good ship, borne steadily on her course by the grateful trade-winds, has passed through the Tropics and reached the bird-latitudes of the South Atlantic, passengers of both sexes, whether berthed in the poop or steerage, welcome with delight the first Albatross. The bright-eyed middy who first descries the snowy breast and mighty wings of the noble bird sharply outlined against the sky, becomes forthwith the central figure of an animated group of enquirers. Very much to his own satisfaction does he pose as a "mariner of the long voyage," as he spins his little yarn of the habits of the great ocean bird, its wonderful powers of flight, and his own pet method of setting about its capture by means of a long line and carefully-baited sail-hook. And very much to the satisfaction of his questioners is it to learn that, now that the tropic zone has been cleared, Cape Pigeon or white-winged Tern, Frigate-bird or Booby, Albatross or Molly-hawk, or some other member of the Ocean Bird tribe, alike under blue skies or lowering rain-clouds, in howling gale or favouring weather, will accompany the ship over some eight thousand miles of water until Australia's shores are sighted.

None but those who have studied Ocean Bird-life from a ship's decks in the course of a long voyage can fully appreciate the charm of the companionship of these beautiful wanderers. The author, who during several passages to and from Australia, India, China, and North America, has derived the greatest enjoyment from such study, has often realised how very little is known by passengers, as a class, of the habits of these birds, and even of the names by which they are recognised by English amateur bird collectors. The following chapters, in which the various Ocean Birds are referred to by their popular seafaring names as used by passengers, as well as by the proper names by which they are known to naturalists, contain descriptions of the plumage and appearance of each bird, which are sufficiently detailed to enable observers to recognise it as it flies near the ship, and also embrace hints as to the best method of capture to be adopted by those desirous of adding specimens to a collection.

The illustrations are from the pencil of Miss Frances E. Green, who has made careful sketches from accurately-stuffed specimens of birds captured by the Author, and I may be allowed to venture the opinion that those who have seen the subjects in their natural state will recognise the fidelity of the drawings.

The Author presents his work to the public in the hope of interesting those bound for distant seas in the winged companions of their travels.

ARTHUR G. GUILLEMARD.

GLOSSARY OF TERMS.

For the following Glossary of Terms employed in this work, and for the accompanying explanatory figure, I am indebted to Dr. F. H. H. Guillemard.

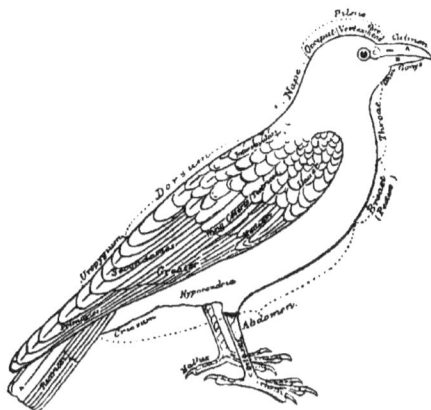

ACUMINATE.—Sharp-pointed.

ALBINO.—"Albinos are animals in which the usual colouring-matters characteristic of the species have not been produced in the skin and its appendages."—*Darwin.*

AURICULARS.—The feathers covering the ears.

AXILLARIES.—The feathers covering the region of the axilla.

CERE.—Bare skin at the base of the bill in certain birds, especially in the Pigeons.

CERVIX.—The nape of the neck.

COMMISSURE.—Line of junction or union.

COVERTS, TAIL (upper).—The feathers sheathing the rectrices of the tail on the upper surface at the posterior part of the uropygium.

COVERTS, TAIL (under).—The corresponding feathers on the under surface, in the region of the vent.

COVERTS, WING (greater and lesser).—Feathers sheathing the remiges, or quill-feathers of the wing (see Figure).

CRISSUM.—The region of the vent.

CULMEN.—The ridge of the maxilla.

DECURVED.—Curved downwards.

FAUNA.—All the animals naturally inhabiting a certain region.

FLORA.—All the plants naturally inhabiting a certain region.

GONYS.—The ridge of the lower mandible.

GULAR POUCH OR SAC.—Dilatation of commencement of alimentary tract, as in the Pelicans.

HALLUX.—The hind toe.

HYPOCHONDRIA.—The side of the body; that part covered by the wings when closed.

IRIS (plural IRIDES).—The coloured portion of the eye surrounding the pupil.

LAMELLATED.—Furnished with lamellæ or little plates.

LAMINATED.—Ensheathed or covered with laminæ or plates.

LORES.—The part between the eye and the base of the bill (c).

c

MANDIBLE.—Term properly applied only to the lower jaw (D); but the *maxilla*, or upper jaw (A), is often called the upper mandible.

MAXILLA.—The upper jaw (A).

MEMBRANE, INTERDIGITAL.—The web between the toes, which in land birds is usually entirely absent, or but slightly marked.

NARES.—The nostrils.

NASAL SHEATH.—Bare skin covering the base of the bill and nostrils, especially marked in the Hawks.

PECTINATE.—Comb-like.

PRIMARIES.—The great flight-feathers of the wing.

RECTRICES.—The quill-feathers of the tail.

REMIGES.—The chief feathers of flight of the wing. They are divided into primaries, secondaries, and tertiaries.

RETICULATED.—Of a net-like aspect.

SCAPULARS.—Feathers of the shoulder region, which cover the side of the back.

SCUTELLÆ.—The horny plates on the anterior aspect of the tarsus.

SECONDARIES.—The secondary flight-feathers of the wing.

SPATULATE.—Spoon-shaped.

SYMPHYSIS.—A growing together or joining of one bone with another.

TARSUS (plural TARSI).—The leg bone immediately above the foot.

TRIDACTYLE.—" Three-fingered, or composed of three movable parts attached to a common base."—*Darwin.*

UROPYGIUM.—The region of and behind the vent.

WEBS (outer).—The feathering or webbing of any of the tail- or flight-feathers on the side away from the median line of the body.

WEBS (inner).—The feathering or webbing of any of the tail- or flight-feathers on the side nearest the median line of the body.

PART I.

PROCELLARIIDÆ.

———◆———

PART I.—PROCELLARIIDÆ.

CHAPTER I.

DIOMEDEINÆ (THE ALBATROSSES).

" And a good south wind sprang up behind,
The Albatross did follow,
And every day, for food or play,
Came to the mariners' halloa."
COLERIDGE.

HE ALBATROSSES (*Diomedeinæ*) are a subfamily and the giants of the great Petrel tribe (*Procellariidæ*). They may always be recognised by their lateral nostrils; as all the rest of the family have them tubular on the ridge of the bill. The two different arrangements are clearly shown in Plate I., the Great Wandering Albatross (*Diomedea exulans*); and in Plate V., the Giant Petrel (*Ossifraga gigantea*).

The Albatross is the monarch of Ocean Birds. No one who has watched one of these birds following a ship in "open water," its broad white breast and enormous spread of wing outlined clearly against the glorious blue of the sky as it soars over the taffrail, will be disposed to question its right to the title. For not only is it the largest in point of size of body, but its extent of wing dwarfs that of all other wanderers on "blue water," and its carriage and flight are stately and imposing to a unique degree.

In an old book entitled 'Grose's Voyage' it is stated that the name Albatross is derived from the fact that the Portuguese called the bird *Alcatraz* (Ostrich), on account of its size, from which word, by corruption, is derived Albatross. Others, again, say that *alcatraz* is the Spanish for a Gannet. So I think it must be admitted that the etymology is doubtful.

B

There are eight distinct species of Albatross now discovered, *viz.* :—

THE GREAT WANDERING ALBATROSS (*Diomedea exulans*, Gould).   Plate I., fig. 1.

THE SHORT-TAILED ALBATROSS (*D. brachyura*, Audubon).   Plate II., fig. 2.

THE CAUTIOUS ALBATROSS (*D. cauta*, Gould).   Plate II., fig. 3.

THE CULMINATED ALBATROSS (*D. culminata*, Gould).   Plate II., fig. 4.

THE YELLOW-BILLED ALBATROSS (*D. chlororhynchos*, Gould).   Plate II., fig. 5.

THE BLACK-BROWED ALBATROSS (*D. melanophrys*, Gould).   Plate II., fig. 6.

THE SOOTY ALBATROSS (*D. fuliginosa*, Gould).   Plate II., fig. 7.

And lastly the newly discovered *D. irrorata*, Salvin, p. 16.

Capt. King, R.N. (Proc. Zool. Soc., 1834, p. 128), says:—"Where one species abounds, the others are only occasionally seen; from which it may be inferred that each species breeds in distinct haunts." This I have noticed to hold good on an entire voyage from England to Australia; so that the times of breeding must be as distinct as the localities.

On account of its superior size, together with its well-earned popularity with seafaring folk, comes first and foremost the GREAT WANDERING ALBATROSS (*D. exulans*), Plate I., fig. 1— *The* Albatross, as it is always considered at sea. This most truly magnificent bird is found chiefly between 30° and 60° S. latitude, and as the famous naturalist Gould tells us, in his 'Australian Birds,' "is constantly engaged in making a circuit of the globe in that particular zone allotted by nature for its habitation." It is occasionally met with in the Tropics as far as 12° South, but when seen there always appears to be completely out of its element. Lord Pembroke and Dr. Kingsley, in 'South Sea Bubbles,' say that an Albatross in a dead calm is one of the meanest of creatures on the wing they had ever seen, and I well remember a long calm in the Tropics with a solitary bird keeping close to the ship, and how laboured his flight appeared. The aspect and carriage of the bird would bear no comparison to that of the dashing-looking specimens seen south of the Cape of Good Hope. Bree includes *D. exulans* in his beautiful 'Birds of Europe.' I have known several instances of their being conveyed across the line in a ship, and let fly on the northern side. Now such birds would be sure to take a northerly course, which probably accounts for their being introduced as an European species.

A distinguishing feature of *Diomedea* is their formidable bill. In *D. exulans* it is pale yellow, and about six inches in length, with the upper mandible hooked at the point and the lower truncated. Like all the Albatrosses the nostrils are large and prominent, and placed widely apart. The head, neck, and body much resemble those of a Goose, both in shape and size, especially when deprived of the extraordinarily thick layer of feathers. The comparatively small size of the bird in that state is well shown at the Natural History Museum (South Kensington), where the skeleton of an Albatross is placed side by side with a stuffed specimen.

The birds vary much in colour, from pure white, speckled or spotted with grey, to a uniform chocolate-brown, and, though a very white bird is of necessity old, and a brown bird is more or less young, this difference is by no means altogether dependent on the age of the bird or the season of the year. The chocolate-brown birds are rather common, and always have a white face. Those I have seen and caught were quite as large as the lighter-coloured birds; in fact the largest Albatross that I know of is a chocolate-coloured bird, now on view at the Natural History Museum.

The specimen from which Fig. 1 was painted I caught in 42° 48′ S., 59° 43′ E. It measured six feet across the wings; and I kept it on account of the unusually beautiful pencil-marking on the breast. All sorts of exaggerated reports are common talk, both at sea and on land, with regard to its spread of wing, which, though enormous, is rarely found to exceed eleven feet four inches from tip to tip. I once made the voyage round the world with a captain who had studied Albatrosses for forty years. During that period he had caught an incredible number, and I myself have frequently seen him capture twelve or so in a day. He always made a point of measuring them most accurately across the wings, and in all his vast experience he never found one over eleven feet four inches. I have caught and measured considerably over a hundred myself, and, curiously enough, my largest birds were also eleven feet four inches from tip to tip.

It is these wonderful wings, with their huge hollow bones, that give to the Albatross its marvellous flying powers. Sailors say that the bird trims its wings to the breeze, and can thus sail along within one or two points of the wind's eye. The action of the wind itself on the wings, however, can have nothing to do with the bird's progression through the air, as it is the resistance of the water that causes a ship to forge ahead when sailing "on a wind." In 'Cassell's Natural History' (vol. iv.), in the description of the Albatross, this matter is cleverly dealt with. Dr. Bennett states that he believes the whole surface of the body of the Albatross is covered by numerous air-cells capable of voluntary inflation or diminution by means of a beautiful muscular apparatus. By this power the birds can raise or depress themselves at will. But there is nothing in their up and down flight sufficiently out of the common to require any apparatus different from other birds, and, with regard to their ordinary flying, Darwin's remarks on the Condor are very applicable. He says:—"The force to keep up the momentum of a body moving in a horizontal plane cannot be great, and this force is all that is wanted—a movement of neck and body appears sufficient."

With respect to this subject, Dr. R. W. Coppinger has the following, in the 'Cruise of the Alert':—"I have had many opportunities of watching the Yellow-billed species (*D. melanophrys*), and I have noticed that it sometimes uses its wings to raise or propel itself in such a manner that to a superficial observer it would then appear to be only soaring with wings stationary. It does not 'flap' them, but depresses them rapidly towards the breast, so that it seems as if the body were being raised at the expense of the wings, whereas, in reality, the entire bird is elevated. The movement does not resemble a flap, simply because the return of the wings to the horizontal position is accomplished by a comparatively slow movement. By resorting to this manœuvre occasionally, it is able to maintain a soaring flight for periods which, without its aid, might be considered extraordinarily long. Of course, when it wants to gain a fresh stock of buoyancy and momentum, it gives three or four flaps like any other bird."

But though it is true that an Albatross cannot soar aloft like an Eagle, and the horizontal flying requires less muscular power, it is in the long, never-wearying flight, carried on for weeks together, day and night without cessation, with a speed sufficient to go with ease in gigantic circles round and round a vessel sailing at a rate of ten or eleven knots, that it stands unrivalled. That such is the case has been often proved by some peculiarly marked bird being observed night after night and day after day following the ship, and is now a well-established fact, although in the old days, when, with close-reefed sails, all hands turned in at night, sailors used unblushingly to affirm that the birds slept out the middle watches on the yard-arms. It is therefore evident that rest, sleep, and drink are alike indifferent to them, and that in their habits they are both diurnal and nocturnal. In a sea-fog they are neither heard nor seen, though if about they would be easily detected. This is then probably one of the few occasions on which they keep to the water; but though the ship may be bowling merrily along the whole time, when the weather clears the same birds will appear, so quickly can they catch up the vessel again.

Their natural food consists of all organic matter the sea may heave up, and principally Squid. The same Captain I have quoted with reference to the size of the birds wrote me, "I have found the mandible of the Great Squid, which is like the beak of a Parrot, and the only hard part of the creature, in the crop of the bird when caught." Likewise, in the examinations made on board H.M.S. 'Challenger,' cuttle-fish (which is the same thing) turned out to be their principal food. In very bad weather they fly with their enormous wings doubled up like a wide-spread **W**—a sort of reefing. It is then that they show their wonderful powers of flying to the best advantage, as with no perceptible effort they forge ahead apparently in the teeth of the very stiffest gale.

The great breeding-place of the Albatross is Tristan da Cunha, a tiny mountainous island in the South Atlantic Ocean. In Moseley's notes on the voyage of the 'Challenger,' describing Tristan da Cunha, is contained the following interesting account of the breeding of the Albatrosses:—"The Albatrosses take up their abode in separate pairs here and there amongst the Penguins or under the trees where there are none of those birds, the latter situation evidently being preferred. The nest is cylindrical in shape, and made of masses of grass and sedge, intermixed with clay. The nest cavity at the top of this solid cylinder is very shallow, and the edge overhangs, through, it is said, the sitting bird picking away the ring of which it is made as it sits. One of these nests measured fourteen inches in diameter, and was ten inches high. They are so compact that when deserted by their owners and the grass has grown over them they make a convenient seat. Only one egg is laid, about the size of that of a goose, or a little larger, elongated, with one end larger than the other. When approached the birds remain sitting quietly or stand by them without attempting to fly away. If disturbed while on their nests they will snap their bills. The male is usually seen with the female at the nest, and marks of affection frequently pass between them. The egg is held in a kind of pouch while it is being incubated, and the bird has to be driven right off its nest before it can be ascertained whether it contains an egg or not. The breeding of these birds takes place in October, November, and December, the summer time of regions lying south of the Equator. The Great Albatross (*D. exulans*) also nests on Tristan da Cunha, within the crater of the highest cone, 7000 feet above sea-level, and their mode of nidification, courtship, &c., is very similar to that of the preceding species." The nests are of course rather larger, and the egg more like that of a Swan than a Goose.

The Albatross is easily caught at sea with a baited hook, the only difficulty being to pay out sufficient line to keep the bait stationary while the ship is forging ahead through the water and the bird is making up its mind to seize the bait. The birds see it directly, as their sight is marvellous, but they have to settle on the water first, and they well know that rising again is a tedious job. In this, however, they are much assisted if there is plenty of wind and a good-sized wave to strike off from. It is then interesting to notice how they run along, rising higher out of the water at each stroke, till they fairly push themselves clear off the undulating surface. To entice a bird down as quickly as possible it is a good plan to throw overboard a handful of scraps, making the bait keep amongst them. The line is floated by means of corks placed close to the hook, which should be a treble one, as there is no time to lose, and the bird must be hooked directly he seizes the bait. Both line and hook should be light, to float as long

c

as possible, and yet very strong, as if well hooked the hauling of the bird on board is simply a question of your tackle holding, and from the fact that you are travelling all the time through the water, no give and take, or "playing" your bird, can be indulged in. Farlow of the Strand, Holroyd of Gracechurch Street, and others, make waterproof cable-laid sea-line in lengths of sixty yards, at about five shillings each, and three such lengths will answer the purpose admirably, while for hooks the largest size treble jack-hooks, such as are used on a spoon-bait, are most suitable. For all the other species of Albatross ("Mollies"), and also the Giant Petrel, exactly the same tackle is necessary, only everything on a slightly smaller scale.

In common with all other fishing, but more particularly here, as the bird is only hooked in the beak, a tight line must be kept. When you commence hauling or winding him in (I always wound up on a winch like a huge log-reel) the bird generally throws back his head, pushes out both his feet, and flaps "hard astern" with his powerful wings. Care must then be taken to prevent his being dragged bodily *under* instead of along the top; as, should this happen the bird gets filled with water, and the weight to pull in is immensely increased. Sometimes, however, he rises in the air, especially in certain latitudes with the wind right aft. In my private log the following notes bear on this subject:—"With a N.W. wind, in and about lat. 13° S. and long. 27° W., on the voyage to Australia,—that is, with the right wind aft,—an Albatross when hooked will fly straight up in the air. This has been noticed for years and years by our sporting Captain (Capt. Austen Cooper, R.N.R., of the 'Carlisle Castle'). It is then a difficult job to bring him on board, as trying to keep a tight line often results in pulling him right over, and he falls on his back in the water, the hook comes out, and—worst of all—the poor bird rarely succeeds in righting himself again."

I remember once seeing an Albatross caught in a very strange manner. It was on a Sunday afternoon on board one of the old Australian sailing-ships. Sunday fishing was prohibited, so of course the birds swarmed, and one Albatross in particular kept hovering right over the man at the wheel. The lead-line was lying out to dry, and the Captain, in a jocular manner, swung it round and round his head and hove it in true nautical fashion, only straight at the bird. The heavy lead took a turn round its wings, and, to the surprise of all hands, including even the Captain himself, the bird half flew and was half dragged on board. He strutted and swaggered about for some time, fighting furiously with all the dogs, and was finally launched into mid-air off the taffrail, as one of these birds is unable to rise from such a smooth surface as a ship's deck.

The first thing an Albatross does on board is to discharge the contents of its stomach. Some say this is a mode of defence; others that he becomes ship-sick. The bird, however, does the same thing when closely pursued by an enemy,—as, for instance, the small but formidable Skua Gull,—a "take this and let me go" sort of policy. Possibly, then, it tries the same tactics when in the power of its still more formidable enemy, man. It next ejects a quantity of oil, of which these birds possess an enormous amount. As they never leave the ocean except for breeding purposes, they must occasionally have to rest in a dangerous sea. Is it possible, then, that they possess the power of ejecting this oil in order to be able to "calm the troubled waters," and therefore, as a last resource in this their new danger, try the same manœuvre that has so often helped them before?

In order to kill an Albatross an extra strong dose of prussic acid should be poured down its throat, and, having tied up its mouth, you may prepare for skinning. Everyone knows the final uses the specimen may then be put to. Its webbed feet make capital tobacco-pouches by drawing out all the bone and leaving on the claws as ornaments. The skin of the feet should be well stuffed with tow, and then pinned out on a board to dry. The wing-bones make excellent pipe-stems; the breast, if carefully cured, a warm though somewhat conspicuous muff; and the beak, in the hands of a skilled artificer, a handsome paper-clip.

In skinning an Albatross,—always a long and tedious job,—too much time cannot be spent in scraping the layers of fat off the skin, and then thoroughly curing with arsenical soap. This being satisfactorily accomplished, at every available opportunity bring the skin up on deck, and leave it in the wind and out of the sun. If these directions are not fully carried out a most disagreeable smell will always remain, and the grease will eventually show through the feathers and spoil the specimen.

Whether an Albatross will attack a person in the water—as, for instance, in the case of anyone falling overboard—is a subject of controversy. The fact is an Albatross swoops down at anything and everything in the sea, from a human being to an old beer-barrel, in all probability out of curiosity, and so indeed do most other sea-birds. Anyone bathing from a ship's boat in "blue water" will find that every bird in his immediate vicinity, both great and small, will, as it were, "go for him"; but though they come uncomfortably near, it is doubtful if any real harm ever ensues. At the same time, the beak of an Albatross is not to be trifled with, and I should not recommend a personal test as a means of solving the question. But it is quite certain that if the bird's object is food it always settles on the water first. This subject is well dealt with by

Lord Pembroke and Dr. Kingsley, who write:—"There seems to be no foundation for the common report that he pounces on his prey, and will with his formidable beak split the head of a man overboard. No bird has less 'picking up' or 'striking power' when on the wing;" and with reference to its settling on the water, the same writers say, "he is as careful of wetting the soft under-feathers of his wings as a lady is of protecting the hem of her petticoat against the mud of the kennel."

The following account, which appeared in the 'Sydney Morning Herald,' has reference to the fact of an Albatross swooping down upon a man who fell overboard from the barque 'Gladstone':—

"On the 24th October, 1881, at noon, whilst the ship was in lat. 42° S. and long. 90° E., and going at the rate of about ten knots an hour, the cry of 'Man overboard!' was raised. Capt. Jackson and his chief officer, Mr. John Rugg, who were seated at dinner at the time, immediately rushed out of the cabin and rounded the ship to. A boat, manned by four hands, was then lowered, and left the ship in charge of Mr. Rugg five minutes after the alarm was raised. The man was then out of sight, but the rescuing party pulled towards the spot where it was supposed he had fallen, and after some little time found him clinging to an Albatross, which he was using as a life-buoy. As soon as the boat got within a few yards of him he let the bird go and swam to the boat, being apparently none the worse for his unexpected immersion. He returned on board smiling, and stated that just after he fell an Albatross swooped down upon him and made a peck at him, but he seized it by the neck and kept its head under water until he had drowned it, and then used it to preserve his own life, in the manner already described. The boat was away about one hour. The sea was very rough at the time, and the wind was from the N.W. The most remarkable thing about this remarkable story is that the man, who could only swim a little, had heavy sea-boots on at the time of the accident, besides being encumbered with oil-skins. The Albatross was the first that had been seen for a month."

The similarity of the Albatross species is so remarkable that in describing the haunts, habits, and characteristics of the Great Albatross we to a very great extent give the leading features of all the species of *Diomedea*. They are, however, generally erroneously divided into three distinct groups by sailors on board an Australian liner:— The Albatross, which is of course *D. exulans*; the Stinkpot, really *D. fuliginosa*; and the Mollyhawk, which embraces all the rest.

There is, however, one species of Albatross, not met with on the Australian voyage, that should be acknowledged by all to be something out of the common, from the fact of its being only found north of the line: for does not the great Dr. Johnson describe an Albatross as a South-Sea bird? This bird (Plate II., fig. 2) is named by Gould the SHORT-TAILED ALBATROSS (*Diomedea brachyura* of Audubon), and by sailors on the China voyages the "China-Sea Albatross," as it is seen more commonly there than elsewhere, but is also found all across the North Pacific to the west coast of North America. It is considerably smaller than *D. exulans*, but is nevertheless a grand-looking bird, measuring some ten feet across the wings. The first I ever saw was in lat. 23° 58′ N., and long. 132° 16′ E.—that is to say, in the China Seas, just out of the Tropics. After a deal of persuasion in the way of ground-bait, it was fairly hooked, but carried away everything owing to the speed of the vessel at the time; but farther north, on other occasions, I was more fortunate. I have noticed that some of these birds have a white streak running round the base of the tail, that gives them the name of "Ringtail" at sea; for it is exactly like the white on the shore-loving "Ringtail," the female Hen Harrier (*Circus cyaneus*). We also considered in this case, too, that they were the females, but I never was fortunate enough to catch one. To sum up on this species, I should say that this is the only Albatross found habitually north of the line; it has, even for an Albatross, a particularly short whitish-marked stumpy tail, and in size is just between the Albatross (*D. exulans*) and the so-called "Mollyhawk" (*D. melanophrys*). Mr. Salvin, in his report on *Procellariidæ* collected by H.M.S. 'Challenger,' says:—

> *Diomedea brachyura.*—Two males and one female. North Pacific. "These were all caught with the hook from the ship while at sea in June and the first half of July, 1874, between Japan and Honolulu; they followed the ship every day in numbers till we got into the trade winds, when no more were observed."
>
> One female. North Pacific. "Eyes brown, bill black; stomach empty. Shot on the 1st April, 1875, by Lord Campbell, with the Henry-rifle, while on the wing. We were just north of the Tropic; but this bird, as well as another Albatross, were seen some days before we had passed out of the Tropics."
>
> One male. North Pacific. "Eyes brown, feet and bill dark or nearly black; stomach had cuttlefish. Caught with a hook, 7th April, 1875." (Proc. Zool. Soc. 1878, p. 740.)

D. nigripes of many writers is only a dark-coloured variety of this bird; in the same way that the chocolate-coloured *D. exulans* is only a variety of the Great Wandering Albatross.

Fig. 3 represents the head of the CAUTIOUS ALBATROSS (*D. cauta*), whose head-quarters are the Bass's Straits. Many of us have often seen this bird, but only considering it an uncommonly wily "Molly" have failed to note it as a distinct species. It is, in fact, as a rule, too clever to be caught, and therefore few opportunities occur for examining it. It may, however, at close quarters always be distinguished from its relatives by the peculiar bright yellow edging at the base of the lower mandible (see Fig. 3). In size it is slightly less than the Short-tailed Albatross, and is nearly allied to, but larger than, *D. melanophrys* (Fig. 6), and is to be met with both on shore and at sea in the same latitudes as *D. exulans*. In Mr. Osbert Salvin's notes to me he says, "Very little known of this bird, except Gould's account." In the 'Proceedings of the Zoological Society,' 1840

(p. 178), the dimensions of a female are given as follows:—Total length, 31 in.; bill, 4½ in.; wing, 21½ in.; tail, 9 in.; tarsi, 3 in. The male is always considerably larger.

Figs. 4 and 5 represent two species so uncommonly alike that it would indeed require a naturalist of the late Mr. Gould's order to have so distinguished them. In mid-ocean they generally both go by the name of the "Golden-beak Molly-hawk."

The specimen from which Fig. 4 is copied is thus summed up in my private log:— "24th Oct. 1867. Caught a Golden-beak Molly, with golden eyes, 6 ft. 9½ in. across the wings. Lat. 42° 05′ S., long. 123° 41′ E." The real name of this bird is the CULMINATED ALBATROSS (*D. culminata*). It is essentially an Australian bird, and is generally found roaming about the S.W. shores of that great continent, but also frequents the Southern, Indian, and South Pacific Oceans.

In Moseley's 'Notes by a Naturalist on the Challenger,' we find the following account, but whether it refers to *D. culminata* or *D. chlororhynchus* (the Yellow-billed Albatross) is not quite clear. Perhaps, like many others, he considers them but varieties of the same bird, and so couples the two names in one:—"The Yellow-billed Albatross (*D. culminata*) breeds on Nightingale Island, the smallest of the Tristan group, about twenty miles S.W. of Tristan da Cunha Island, and about a square mile in extent. The whole of this island, except the steepest slopes and the highest peaks, about one thousand feet above the sea, is covered with a luxuriant growth of tall grass, almost impenetrable, higher than a man's head, and studded here and there with clumps of trees. This sea of verdure is intersected by a long lane or street through which the Penguins make their way to the sea."

The specimen from which Fig. 5 was drawn was caught by me in lat. 54° 16′ S., long. 112° 53′ W., and in my private log is described as the Blue-necked Golden-beaked Molly.

Now both *D. culminata* and *D. chlororhynchos* are blue-necked and both are golden-beaked, and they are no doubt closely allied, but I should be inclined to say that *D. culminata* is a larger and heavier-looking bird, especially about the neck, where there is also less blue than in *D. chlororhynchos*. The eyebrow is also much less distinctly marked. There is also a considerable difference in the colouring of the beak and in the way in which the golden-yellow is laid on, which is clearly shown in the illustrations. The webbed feet of *D. culminata* are also proportionately larger than those of *D. chlororhynchos*. The largest specimens I ever caught of either measured eight feet across the wings. Gould says *D. chlororhynchos* fairly dives after its prey, swimming under the water the while.

Fig. 6 represents the BLACK-BROWED ALBATROSS (*Diomedea melanophrys*), our dear old friend the common Molly-hawk, Molly-mauk, or Molly. This is by far the commonest and most sociable of all the family, and except for the fact that you may be seeing the very same birds day after day, as they all appear so exactly alike, one might think the Southern Hemisphere produced these birds in extraordinary numbers. The specimens met with at sea are nearly always almost identical both in size and colouring, thereby greatly differing from the Great Wandering Albatross with its extraordinary variations. It is very rarely seen without the well-defined black band across the back and wings (which is pointed out as a distinguishing mark of the Molly to passengers on board ship); the snowy-white head, breast, and body; the bright yellow beak; and, lastly, the dark-coloured eyebrows, from which it derives its name. There is no visible difference in the sexes, but those who have seen them on their breeding-stations say the younger the bird the browner the bill. It is the most easily caught of all the Albatrosses, as it comes down quicker and takes a bait less leisurely than the others of the *Diomedea*. The largest specimen I ever caught was in lat. 38° 59′ S., long. 143° 33′ E., which measured eight feet and an inch across the wings. In 1885 a fine specimen was obtained for the Zoological Gardens, concerning which I addressed the following letter to the Editor of ‘The Times,’ which appeared on December 5th, 1885:—

AN ALBATROSS IN THE ZOOLOGICAL GARDENS.

Sir,—A few months ago you permitted me to write in ‘The Times’ about a giant Petrel, misnamed by some an Albatross, that was deposited for a short time in the Zoological Gardens by Mr. Jamrach. Thanks to Mr. Ayshford Sandford, there is now in the Eastern Aviary, facing the cage of the pretender, a magnificent Black-eyebrowed Albatross (*Diomedea melanophrys*).

This is, I think, the only Albatross that has ever reached our shores alive; but, now that the feat has been so successfully accomplished, there seems no reason why we should not also one day possess a specimen of the Great Wandering Albatross (*Diomedea exulans*), the largest by far of all the Petrel tribe, and one of the grandest birds in the world.

Those that ‘go down to the sea in ships’ will recognise in this new arrival the sociable and popular Mollyhauk, Mollymauk, or Molly, so easily distinguished by the broad black band across its back.

In a large open grass aviary, with a fine stream to swim in, a huge grass mound to roost on, unlimited fresh fish to eat, and a young Gannet to bully and chum with, this bird is doing well.

I remain your obedient servant,

Blackwall Yard. JOS. F. GREEN.

In the Society of Acclimatisation at Sydney they have or had both *D. exulans* and *D. melanophrys* in captivity; but that is, of course, a very different undertaking, as the birds are close at hand.

Fig. 7 represents the Sooty Albatross (*Diomedea fuliginosa*), commonly called at sea the "Stinkpot"—a name which, with all respect be it spoken, might be applied to the whole class. This Albatross differs in appearance from all its tribe, and no one would imagine that it is of the same family as the grand-looking birds described above. It is of much slimmer build, and it is doubtless from this fact that the late Mr. Gould said, "the unrivalled flight of this Albatross carries off the palm from all competitors." Its plumage is sooty-black all over; eyes golden, and jet-black beak and feet. Its great object at sea appears to be to study minutely the trucks of the masts, especially if vanes are carried, round which it sails in the most marvellous manner, keeping beautifully even pace with the vessel the while. This over-curiosity on its part is tempting to passengers anxious to try the merits of their newly-bought fire-arms. This, however, as might be imagined, would be likely to lead to serious consequences to the rigging; so in the ship on which the specimen was shot from which Fig. 7 was drawn, only rifles were allowed, and that at stated times, whilst only one person was permitted to fire at a time. The birds themselves took absolutely no notice of the proceedings unless hard hit, and it is a curious fact that no amount of noise or close whirr of bullets appears to have the slightest effect on an Albatross. I shot this particular bird in lat. 40° 05′ S., long. 3° 11′ W., and a more unpleasant job than the skinning of it cannot well be imagined. With respect to this bird, Moseley says:—"The Sooty Albatross (*D. fuliginosa*), called 'Prew' or 'Pro' by the sealers, breeds on Marion Island, and does not appear to nest low down like other species." Mr. Gould notes its first appearance in July, in one of his voyages, in lat. 31° S.

The eighth species of Albatross I have dealt with at the end of this Chapter; and as I have never seen the bird there is no illustration given.

The generic characters of the foregoing, as given by Gould, in his magnificent work 'The Birds of Australia,' will greatly assist the seafarer to determine the different species. The description of *Diomedea irrorata* was given me by Mr. Salvin, who possesses the only specimen in this country:—

1. The Wandering Albatross (*Diomedea exulans*).—"Varies much in colour at different ages; very old birds are entirely white, with the exception of the pinions, which are black; and they are to be met with in every stage from pure white, white freckled and barred with dark brown, to dark chocolate-brown approaching to black, the latter colouring being always accompanied by a white face, which in some specimens is washed with buff; beneath the true feathers they are abundantly supplied with a fine white down; the bill is delicate pinky white, inclining to yellow at the tip; irides very dark brown; eyelash bare, fleshy, and of a pale green; legs, feet and webs pinky white. The young are

at first clothed in a pure white down, which gives place to the dark brown colouring mentioned above."

2. The Short-tailed Albatross (*Diomedea brachyura*).—"The adults of both sexes have the general plumage white, washed with buff on the head and neck; the edge and centre of the wing white, the remainder and the tips of the tail dark brown; bill pinky flesh-colour; irides brown; legs and feet bluish white; eyelash greenish white. The young differ in being of a uniform chocolate-brown."

3. The Cautious Albatross (*Diomedea cauta*).—"The beautiful grey on the sides of the mandibles, the delicate pale yellow of the culmen, and the yellow mark at the base of the lower mandible will at all times distinguish this bird from the other members of the genus. Crown of the head, back of the neck, throat, all the upper surface, rump and upper tail-coverts pure white; lores and line over the eye greyish black, gradually passing into the delicate pearl-grey which extends over the face; back, wings, and tail greyish brown; irides dark vinous-orange; bill light vinous-grey, or bluish horn-colour, except on the culmen, where it is more yellow, particularly at the base; the upper mandible surrounded at the base by a narrow belt of black, which also extends on each side of the culmen to the nostrils; base of the lower mandible surrounded by a belt of rich orange, which extends to the corners of the mouth; feet bluish white; irides brown. When fully adult the sexes differ but little in colour; the female may, however, at all times be distinguished by her diminutive size, and the young by the bill being dark grey."

4. The Culminated Albatross (*Diomedea culminata*).—"Back, wings, and tail dark greyish black, the latter with white shafts; head and neck white, washed with greyish black; round the eye a mark of greyish black, interrupted by a streak of white immediately below the lower part of the lid; rump, upper tail-coverts, and all the under surface pure white; bill black; the culmen horn-colour; and the edge of the basal three-fourths of the edge of the upper mandible orange. In the youthful state the head and neck are dark grey, and the bill is of an uniform brownish black, with only an indication of the lighter colour of the culmen."

5. The Yellow-billed Albatross (*Diomedea chlororhynchos*).—"Spot before a line above the eye washed with grey; head, neck, all the under surface, rump, upper tail-coverts, and under surface of the wing snow-white; back and wings brownish black; tail brownish slate-colour, with white shafts; culmen from near the base to the point bright orange-yellow; remainder of the bill black; irides greyish brown; feet bluish white."

6. The Black-eyebrowed Albatross (*Diomedea melanophrys*).—"Head, back of the neck, all the under surface, and the upper tail-coverts pure white; before, above and behind the eye a streak of blackish grey; wings dark brown; centre of the back slaty black, into which the white of the back of the neck gradually passes; tail dark grey, with white shafts; bill buffy yellow, with a narrow line of black round the base; legs and toes yellowish white, the interdigital membrane and the joints washed with pale blue; irides very pale brown, freckled with a darker tint."

E

7. THE SOOTY ALBATROSS (*Diomedea fuliginosa*). — "The whole of the plumage deep sooty grey, darkest on the face, wings, and tail; shafts of the primaries and tail-feathers white; eyes very dark greyish brown, surrounded, except anteriorly, by a beautiful mark of white: bill jet-black, with a longitudinal line of white along the under mandibles, this white portion not being horny like the rest of the bill, but composed of fleshy cartilage, which becomes nearly black soon after death; feet white, slightly tinged with fleshy purple."

8. *Diomedea irrorata*. — Capt. Markham, of Arctic renown, was fortunate enough to discover this new species of Albatross in Callao Bay. The bird is now in the possession of Mr. Osbert Salvin, and was thus described and named by him in the 'Proceedings of the Zoological Society of London' of June 15th, 1883 : — "*Diomedea irrorata*, sp. n. Male. Callao Bay, Peru, December, 1881. This Albatross appears to be quite distinct from any hitherto known. It appears to come next to *D. melanophrys*, having the bill similarly constructed (*cf.* Cones, Pr. Ac. Phil. 1866, pp. 186, 187), but the bill is much longer and the bird larger in all its dimensions, except the tail, which is shorter and more rounded. In coloration, too, there is great difference, the upper back and rump being variegated with dusky and white instead of pure white, and the abdomen wholly dusky with minute white freckles." Doubtless *D. irrorata* is descended from some well-known members of the family who, for some reason or other, took up their abode in this tropical part, and gradually adopted a plumage and structure to suit their new surroundings. The present stage of the bird is probably of recent date, as in such a thoroughfare as Callao Bay so large a new species could hardly have existed unobserved for any length of time.

So it would seem that the *Diomedeinæ*, as at present known, do not necessarily represent the whole of the family, and even now the dark-coloured *D. nigripes* is considered by many to be a distinct species, and not the young, or a variety, of *D. brachyura*. Mr. Swinhoe also names five very black Albatrosses he obtained in China as *D. derogata*, making a third species found north of the Line. But I think most of our great ornithologists consider both *nigripes* and *derogata* the same as *brachyura*, and I humbly follow the majority.

CHAPTER II.

THE SMALL PETRELS.

" She swept the seas; and as she skimmed along,
Her flying feet unbathed on billows hung."

DRYDEN.

AND so with our tiny little, well-known ocean wanderer, the Stormy Petrel (*Procellaria pelagica*); and hence, as referring to the power and habit of apparently walking on the surface of the water, the name Petrel, which is common to the whole family, and derived from the Apostle Peter, who walked on the water. Some years ago Gould and others named the birds described in this Chapter *Thalassidroma* (θαλασσα and δρομη), as descriptive of birds "running" in the "sea." But of late this has been abandoned; and I have now only coupled them together, as being to the sailor simply "Mother Carey's Chickens," a name said to have been bestowed upon them by Captain Carteret's sailors, but for what reason does not appear to be known. I have heard that it means the "Mother carries her Chickens," from the fact of her being so continually on the wing; but this we must all allow is rather far-fetched. Indeed the name seems wrapped in mystery.

The Stormy Petrel is the smallest web-footed bird known. Gould says, "Assuming that the Great Albatross usually weighs about fifteen pounds, and the Storm Petrel an ounce, the former is 240 times as heavy as the latter." The meaning of *pelagica* is "of or belonging to the sea," and *Procellaria* (of which this bird is now the recognised type), from *procella*, "a storm"; both of which terms are truly applicable to this little wanderer, who is to be met with in nearly every sea in the Northern Hemisphere, and equally at home in any weather. He is thus described by Yarrell:—"The bill is black; the irides dark brown; head, neck, back, wings, and tail sooty-black; outer edges of tertials white; upper tail-coverts white; chin, throat, breast, belly, and under tail-coverts sooty-black; legs, toes, and membranes black. The whole length of the bird not quite six inches; the wing, from the bend, four inches and five-eighths. The young bird, till twelve months old, is not quite so dark in colour; edges of wing-coverts rusty brown, and no white on the margins of the tertials."

The Stormy Petrel is common all round the shores of Great Britain, though rarely seen except by sailors. Occasionally, however, they are driven in shore by heavy gales of wind, and picked up in a miserably exhausted condition. Dresser says:—"This inhabitant of the ocean, appearing only to occur about the land during the breeding season and when driven in by stress of weather, has a tolerably extensive range, being found throughout the Atlantic Ocean, and having also been met with on the east coast of Africa."

In the winter of 1882 I saw the specimen from which Plate IV. is painted, feebly flapping along the Thames off Greenwich Pier, and with the aid of a waterman succeeded in capturing it.* The poor little thing lived some days by sucking its feathers, which I had plentifully besprinkled with oil, but finally succumbed, and is now in my collection of British birds.

That charming naturalist Charles Waterton says, in his explanatory index, that the Stormy Petrel is "too well known to need description." He, however, proves the fallacy of this off-hand treatment of the subject by mentioning, in his 'Wanderings,' that it is only seen when a heavy gale is blowing, and that "when the storm is over it appears no more." Now, though no bird is more thoroughly at home in bad weather than a "Mother Carey," yet it is by no means consistent with facts to say it is only seen then. Indeed at sea the opposite is rather the case, for in rough weather the Crustacea and Mollusca, and other minute organisms, upon which the birds feed, are brought to the surface by the action of the waves.

Now this friendly office of turning up food is also performed by the moving ship, especially in the case of a paddle or screw, and naturally more appreciated when the sea is smooth, and there is no other way of obtaining these submerged delicacies. It is also then that the host of eagerly following birds are so keenly on the look-out for all scraps and refuse that are thrown them, or fall from the ship itself. When a vessel is becalmed or at anchor I have seen them settle in flocks alongside, prepared for a regular square meal off anything they can get. Referring to their rapacity, Yarrell says:—"On examining the inside of a Stormy Petrel, Mr. Couch found about half an inch of a common tallow candle, of a size so disproportionate to the bill and throat of the bird that it seemed wonderful how it could have been able to swallow it." Thus, then, it stands to reason the finer the weather the more dependent are the birds on the inventions of man for their daily food, and consequently the more seen by them. I am

In March, 1886, the same Greenwich waterman handed me a live Red-legged Partridge that he found swimming about off the pier.

therefore, with all modesty, compelled to disagree with Waterton, and all who affirm that this little Petrel is only found in stormy weather.

There is, however, no denying the fact that by many mariners the Stormy Petrel is considered the sure harbinger of storms and calamities, and by such looked upon as a bird of ill-omen. Thus the name of Stormy Petrel, or "Stormfogel" of Northern Europe ; and *Procellaria*, from *procella*, "a storm."

Especially was this the case in the old superstitious days, when, in addition, the birds were looked upon as a sort of repository for the souls of departed seamen, giving to them a knowledge on storm-lore unattainable by other birds, but considered of so ghostly a character that no practical use appears to have been made of it by the sailor.

> "Outflying the blast and the drifting rain,
> The Petrel telleth her tale in vain ;
> For the mariner curseth the warning bird
> Who bringeth him news of the storms unheard."
>
> BARRY CORNWALL.

Now doubtless the manner of all birds will foretell the coming storm to those able to read the signs aright ; for as Davenport Adam says :—

> "From birds, in sailing, men instruction take ;
> Now lie in port, now sail and profit make."

And amongst them must, of course, be included the "Mother Carey ;" but surely nothing uncanny should be attributed to this the most vivacious, the most contented little bird on the whole wide ocean. Food, weather, and general surroundings may be as contrary to their natural tastes as possible, but they will appear as cheery as ever.

This contented disposition struck me most forcibly in the poor little half-starved specimen I caught at Greenwich. Imagine it prostrate on flannel, thoroughly exhausted, and dripping with oil (what an insult to a "Mother Carey," who gives us the purest oil imaginable) ; and yet there it lay sucking the make-shift off its feathers without so much as a grimace, and between whiles singing a sweet little warble, suggestive of the purring of a well-fed petted tabby. How easily pleased ; Mark Tapley himself could not have behaved in a more exemplary manner.

I feel sure that in their own bird-world they never grumble ; and this in itself, by-the-bye, should for ever free them from the imputation of being possessed of the souls of departed sailors ; for who ever heard of a Jack Tar that did not occasionally indulge in a growl—a sailor's privilege all the world over. And how our little friendly Petrel loves a game : who has not seen them at sea racing the skipjacks for the pure fun of

F

the thing, apparently playing a sort of "touched you last" with the astonished fish! Let us, then, dismiss all superstitions detrimental to them from our minds; and rather be it our pleasure to minutely study the birds, and thereby interpret for our own benefit such prophetic signs as their habits may be capable of giving us. Wilson says on the subject:—"As well might they curse the midnight lighthouse that, star-like, guides them on their watery way, or the buoy that warns them of the sunken rock below, as this homeless wanderer, whose manner informs them of the approach of the storm, and thereby enables them to prepare for it." Happily, however, they are comparatively safe when following a ship at sea. Be it from love or be it from fear, the result to them is the same; most sailors will protect the "Mother Carey's Chicken."

I well remember some years ago seeing a quartermaster leave his wheel, and forcibly take the line out of a passenger's hand to free a little "Mother Carey" entangled therein. I knew the sailor well; and waves running mountains high, threatening to engulf him and poop the ship, would not have induced him to leave the helm, for a first-rate sailor was he. Taciturn and respectful, too, and by no means given to insulting passengers; but this enormity carried out before his very eyes was more than he could stand. The Captain soon heard of it, but wisely considered it a misdemeanour best ignored, and at the same time, with much tact, smoothed down the ruffled feelings of the indignant fisherman. To those, then, anxious for specimens, and on board vessels where such feelings exist, I would say do your fishing at night, or unobserved, and by this means you will avoid openly wounding the susceptibilities of any sensitive or superstitious mariner.

I was fortunate in one voyage in being the possessor of a stern cabin, and many is the "Mother Carey" I have hauled in at the port to keep as a specimen, or examine and let fly again, as they are not in the slightest degree injured by the process. All the gear required is thread with a cork at the end, rounded so as to avoid any jerky strain from the waves. To obtain the right shape, partially burn the cork that you have already cut into the appearance of a large marble, and then rub it round and round in your hands. The birds fly against the line and entangle themselves. I have invariably found you could catch twice as many fishing at night; sometimes, of course, a Cape Pigeon or other large Petrel carries away all your gear, but that is soon rectified. Dr. Coppinger, in his 'Cruise of the Alert,' makes a strong point of the fact to prove they are on the wing all night.

On shore the treatment of these charming little birds is of a very common-place order. Morris tells us that the inhabitants of the Ferroe and other islands use them for lamps. A wick of cotton or other material is drawn through the body, which when lighted continues to burn till the oil in the body is consumed. In these same islands they are often caught (and afterwards released) for the sake of the valuable oil, which, like all the Petrels, they vomit on being handled.

The Stormy Petrel is noted as being the latest layer all round our shores. It makes a nest of *debris* in any hole on the ground, where it deposits one white oval egg, about an inch long.

The Rev. J. G. Wood says : — "Who would think, on inspecting a specimen of the well-known Stormy Petrel, that it was able to dig into the ground, and form the burrow in which it makes its nest? Such, however, is the case; and the pretty little traverser of the ocean shows itself to be as accomplished in excavating the ground as it is in flitting over the waves, with its curious mixture of flight and running. If the Stormy Petrel can find a burrow already dug it will make use of it, and accordingly is fond of haunting rocky coasts, and of depositing its eggs in some suitable clefts. It also will settle in a deserted rabbit-burrow, if it can find one sufficiently near the sea; and is found breeding in many places which would equally suit the Puffin. Failing, however, all natural or ready-made cavities, the Stormy Petrel is obliged to excavate a tunnel for itself, and even on sandy ground is able to make its own domicile. Off Cape Sable, in Nova Scotia, there are many low-lying islands, the upper parts of which are of a sandy nature, and the lower composed chiefly of mud. Not a hope is there in such existing localities of already existing cavities, and yet to those islands the Petrels resort by thousands, for the purpose of breeding. The birds set resolutely to work, and delve little burrows into the sandy soil, seldom digging deeper than a foot, and in fact only making the cavity sufficiently large to conceal themselves and their treasure. Each bird lays a single egg, which is white, and of small dimensions. The young are funny-looking objects, and resemble puffs of white down rather than nestlings. The parent attends to its young with great assiduity, feeding it with the oleaginous fluid which is secreted in such quantities by the digestive organs of this bird."

WILSON'S STORM PETREL (*Oceanites oceanicus*).—This is the commonest of the ocean "Mother Careys," and is met with in equal numbers on both sides of the Equator. In the voyage to Australia we should take them with us from the Land's End to the Tropics, and again from the south side of the Tropics all the way to Australia. Gould says it is the only Petrel found on both sides of the line. In size and appearance it is much like the Stormy Petrel, but is slightly larger and leggier. Dr. Coppinger, in the 'Cruise of the Alert,' remarks that in the Pacific the Storm Petrels are in the habit of kicking the water with one leg while skimming the surface in search of food. This we have doubtless all observed. He then goes on to say that the Atlantic Storm Petrels "steady themselves on the water with both legs together," instead of giving this one-legged kick. Perhaps the interpretation of this is that they are two different birds, the Pacific Storm Petrel being Wilson's Storm Petrel, and the Atlantic the common Stormy Petrel. It is almost impossible to tell them apart, at any distance from the ship; but, having the bird in your hand, it is as impossible to confuse them together, from the fact of Wilson's having yellow patches on the webs of the feet. The bird is thus defined by Yarrell:—"The bill is black; the irides dark brown; the head, neck, back, wing-primaries, and the tail-feathers dark brownish black; greater wing-coverts and the secondaries dark rusty brown, lighter in colour near the end, with the extreme edges and tips white; upper tail-coverts white; chin, throat, breast, and all the under parts sooty black, except some of the under tail-coverts, which are tipped with white; legs long and slender, with the toes and their membranes black, but with an oblong greyish yellow patch upon each web." In Ornithology this bird is the only representative of the genus *Oceanites*. Some collectors mention the under tail-coverts black instead of white-tipped. My specimens have them quite white, and Gould colours them so in his 'Birds of Australia.'

THE FORKED-TAILED PETREL (*Cymochorea leucorrhoa*).—All "Mother Careys" remind me of Swallows, but this one particularly, by reason of its forked tail, by which it is always to be recognised. Describing this bird Audubon says:—"The species of this genus, with which I am acquainted, all ramble over the seas, both by night and by day, until the breeding-season commences; then they remain in their burrows, under rocks, or in their fissures, until towards sunset, when they start off in search of food, returning to their mates or young in the morning, and then feeding them. When you pass close to the rocks in which they are you can easily hear their shrill, querulous notes; but the report of a gun silences them at once, and induces those on the ledges to betake themselves to their holes. The Forked-tailed Petrel emits its notes night and day, and at not very long intervals, although it is less noisy than Wilson's Petrel. They resemble the syllables 'pour-wit,' 'pour-wit.' Its flight differs from that of the other two species, it being performed in broader wheelings, and with firmer flappings. It is more shy than the other species; and when it wheels off, after having approached the stern of a ship, its wanderings are much more extended before it returns. I have never seen it fly close around a vessel as the others are in the habit of doing, especially at the approach of night; nor do I think that it ever alights on the rigging of ships, but spends the hours of darkness either on the water or on low rocks or islands. It also less frequently alights on the water or pats it with its feet; probably on account of the shortness of its legs, although it frequently allows them to hang down. In this it resembles the Storm Petrel, and Wilson's Petrel has a similar habit during calm weather. I have seen all the three species immerse their heads into the water to seize their food, and sometimes keep it longer under than I had expected. The Forked-tailed Petrel, like the other species, feeds chiefly on floating Mollusca, small fishes, Crustacea, which they pick up among the floating sea-weeds, and greasy substances, which they occasionally find around fishing-boats or ships out at sea. When seized in the hand it ejects an oily fluid through the tubular nostrils, and sometimes disgorges a quantity of food. I could not prevail on any of those which I had caught to take food." Dresser says:—"This bird, so essentially a bird of the ocean, has, as may be supposed, a tolerably extensive range, being found in the Atlantic from St. Kilda and the coast of Labrador, southward on the American coast to Washington, and on our side to Madeira." The bird is thus described by Yarrell:—"The bill is black; the irides dark brown; the head, neck, and back sooty black, the back rather the darkest in colour; wing-coverts rusty brown; the tertials tipped with white; upper tail-coverts white; primaries and tail-feathers black; the tail forked, the outer feathers being half an inch longer than those in the middle; breast sooty black; behind each thigh and extending to lateral under tail-coverts an elongated patch of white; the middle under tail-coverts sooty black. The whole length of my bird, seven inches and a quarter; from the anterior bend of the wing to the end, six inches. The sexes in plumage are alike."

BULWER'S PETREL (*Bulweria columbina*).—On the Australian voyage this Petrel may be looked for about Madeira and the adjacent western coast of Africa. Dresser says:—"The present species is restricted entirely to the Atlantic Ocean, being met with chiefly on or near the Canaries and Madeira." It is thus described by Yarrell:—"The bill is black; the irides nearly so; the whole of the plumage almost sooty black, rather paler on the edges of the great wing-coverts; tail rounded; legs and toes dark reddish brown, the interdigital membranes dark brown. The whole length, from the point of the beak to the end of the tail, ten inches and a half."

WHITE-FACED STORM PETREL (*Oceanodroma marina*). — Thus described by Gould:— "Forehead, face, line over the eye, and all the under surface pure white; crown and

G

nape, a broad patch beneath the eye, and the ear-coverts, slate-colour; sides of the chest, back of the neck, and upper part of the back, dark grey, gradually passing into the dark brown of the back and wings; upper tail-coverts light grey; primaries and tail black; irides dark reddish brown; legs and feet black; webs yellow." In the 'Proceedings of the Zoological Society' for 1878 (p. 736), Mr. Salvin, describing some specimens, says:— " Nightingale Island. Eyes black; a night-bird. These were taken out of holes in the ground during the day by help of the dogs."

BLACK-DELLIED STORM PETREL (*Fregeta melanogaster*). — Thus described by Gould:— " All the plumage deep sooty black, with the exception of the upper tail-coverts and flanks, which are snow-white ; bill, legs and feet black." In the voyage of H. M. S. ' Challenger ' one was observed in Betsy Cove, Kerguelen ; and in Lord Lindsey's Expedition Mr. Saunders mentions their being caught in lat. 36° 57' S., long. 40° 41' E.

WHITE-DELLIED STORM PETREL (*Fregeta leucogaster*).—Thus described by Gould:—" Head and neck deep sooty black; back greyish black, each feather margined with white; wings and tail black; chest, all the under surface, and the upper tail-coverts white; bill and feet jet-black." Thus described by Mr. Salvin in his report on the collection of H. M. S. ' Challenger ':—" Eyes brown. Shot in the South Pacific, 11th November, 1875, at sea. Their stomachs were filled with a yellow oil, and mixed with it some pieces of Crustacea."

GREY-DACKED STORM PETREL (*Garodia Nereis*).—Thus described by Gould :—" Head, neck, and chest sooty grey; lower part of the wing-coverts, back, and upper tail-coverts grey, each feather very slightly margined with white; wings greyish black; tail grey, broadly tipped with black ; under surface pure white ; irides, bill, and feet black." In the Proc. Zool. Soc. for 1840, p. 178, he also says:—" Total length, six inches and a half; bill, nine-sixteenths of an inch ; wing, five inches and a quarter ; tail, two inches and a half ; tarsi, one inch and a quarter. Hab.—Bass's Straits, on the south coast of Australia. This beautiful fairy-like Storm Petrel is about the size of *Wilsoni*, and is remarkable as differing from most of the members of the group in having no white on the rump and in the pure white of the under surface." In March, 1858, a dead specimen was picked up in the Falkland Islands.

CHAPTER III.

OTHER SPECIES OF PETREL.

" Up and down! Up and down!
From the base of the wave to the billow's crown,
And amidst the dashing and feathery foam
The daring Petrel finds a home,—
A home, if such a place may be,
For her who lives on the wide, wide sea,
On the craggy ice, in the frozen air,
And only seeketh her rocky lair
To warn her young, and to teach them spring
At once o'er the waves on their stormy wing!"

BARRY CORNWALL.

IN the two preceding Chapters all the smallest, and with one exception the largest, Petrels have been dealt with. This third Chapter will comprise all the remainder that we should be likely to fall in with on the Australian voyage.

THE GIANT PETREL (*Ossifraga gigantea*), of the genus *Ossifraga*, and the only species known, is a familiar bird to all traversers of the Southern Seas. As the name implies, it is a bird of huge proportions, equalling in general dimensions the smaller Albatrosses, and, as Prof. Huxley says, "holds a sort of middle place between the Gulls and the Albatrosses." Its *nom de mer* is "Nelly," applied to either sex ; it is also called Leopard-bird, or Leopard Albatross. It is often considered an Albatross both at sea and on land ; but the curiously built-up beak, with nostrils encased in one sheath (instead of on either side), should at once show it to be no *Diomedea*. The flight of this great Petrel is, moreover, inferior to that of the Albatross family, having considerably more of the flapping land-bird style about it, by which it can always be recognised at a distance. This inferiority, however, is only by comparison, as I have often noticed a particular specimen follow the ship for days together, and then only leave by reason of its being caught with hook and line. If they are on the feed, this is easily accomplished with your Albatross fishing-gear ; and I have frequently caught two or three in a day, though, as a rule, they are shyer than the *Diomedea*. I caught the specimen from which Fig. 5 was copied in lat. 42·05° S., long. 123·41° E., and it measured eight feet across the wings. The Giant Petrel is extremely powerful, and also excessively fierce and predacious—a combination most disastrous to the smaller Petrel tribe, off whom it loves to make a meal. The very name of *Ossifraga* (Breakbones) has a fearfully carnivorous ring about it, and indeed there is no doubt that any fish, flesh, or fowl of moderate dimensions becomes an acceptable addition to the daily meal of this gigantic Petrel. Supposing it to be the fact that a so-called Albatross has been seen to attack a man in the water, I should imagine that the *Diomedea* tribe was unjustly suffering for the sins of a bad-tempered or half-starved *Ossifraga*. In Gould's 'Birds of Australia' the following account appears, quoted from 'The Ibis' for 1865:—"Capt. F. W. Hutton states that the bird (Giant Petrel) breeds in the cliffs of Prince Edward Islands and Kerguelen's Land, but the nest can be got at occasionally. The young are at first covered with a beautiful long light grey down; when fledged they are dark brown, mottled with white. When a person approaches the nest the old birds keep a short distance away, while the young ones squirt a horridly smelling oil out of their mouths to the distance of six or eight feet. It is very voracious, hovering over the sealers when engaged cutting up a Seal, and devouring the carcase the moment it is left, which the Albatross never does. It sometimes chases the smaller species, but whether or not it can catch birds possessed apparently of powers of flight superior to its own is doubtful; but, supposing one killed, that it feeds only on its heart and liver I cannot believe; yet it is said to do so in the words of many ornithologists." Gould says Capt. Cook found it very abundant on Christmas Island, Kerguelen's Land, and so tame that the sailors knocked them down with sticks. The entire plumage of the adult bird is chocolate-brown; bill pale straw-colour; irides and legs dark brown. In immature plumage they are spotted with white—hence the name "Leopard-bird." The pure white (albino) varieties are by no means uncommon. There is an excellent specimen in the Natural History Museum; also a very pale buff-coloured one, besides the usual chocolate-coloured bird. Mr. H. Saunders, on the sea-birds collected by Lord Lindsay's Expedition, says:—"*Ossifraga gigantea*, 'Cape-hen' (*sic*), No. 50, male, Sept. 10th, lat. 34° S., long. 10° 42′ W. Beak pale apple-green, much darker at the tip; iris dark brown; feet sooty black." No. 52, Sept. 14th. "Beak greyish green, darker at tip; iris dark brown; feet silvery brown; spread of wing, six feet seven inches." In 1885 I received a letter from Mr. Jamrach, stating that he had deposited a species of Albatross in the Zoological Gardens; on hastening to see this wonderful addition to their aviaries I found a fine specimen of the Giant Petrel, though going by the name of Short-tailed Albatross. It only stayed a few days, as arrangements had been made for sending it to Paris. It appeared in good health, and fed voraciously off fresh herrings. This is, I think, the only specimen that has reached this country alive. I sent the following short notice of the event to the 'Times,' which appeared June 8th, 1885:—

AN ACQUISITION TO THE ZOOLOGICAL GARDENS.

Sir,—I received a letter the other day from Mr. A. H. Jamrach, the well-known naturalist dealer, saying he had deposited a live Albatross in the Zoological Gardens. On receipt of this startling news I hastened off at once to inspect.

I think, however, I am right in saying that the bird is not an Albatross (*Diomedea*), but the Giant Petrel (*Ossifraga gigantea*), well known at sea, but unquestionably a *rara avis* in these climes, indeed quite as much so as the true Albatross. In size and extent of wing they are about the same as the smaller species of *Diomedea*, but in no way approach the great Wandering Albatross (*Diomedea exulans*), with its enormous spread of eleven feet from tip to tip. Their flight is much the same, but this Petrol has more of the land-bird flap about it, by which it can be recognised at a great distance.

This new arrival appears fairly well, and fed heartily off some fresh herrings; but I cannot help thinking it would thrive far better in the fine open aviary opposite its present cage, and there would be no fear of its escaping from such an enclosure. They are often kept for days together on board ship, and are quite unable to fly, unless taken up and launched into mid-air off the rail.

I am your obedient servant,

Blackwall Yard.   JOS. F. GREEN.

The word "Ossifrage" (Heb. *peres*, γρψ, *gryps*) occurs twice in Holy Scripture, as a bird that may not be eaten: in Lev. xi. 13, and in the parallel passage in Deut. xiv. 12; but it is not probable that this particular bird was meant. In the new version the word is omitted altogether.

GREAT GREY PETREL (*Adamastor cinerea*).—This bird (well known at sea as the Cape Dove) is sometimes called the Capped Petrel. It is a combination of *Procellaria* and *Puffinus*, and is one of the Southern Seas representatives of the well-known Shearwater family. On my sending up a specimen that I had caught off the Cape to one of the very greatest and most obliging of our ornithological authorities, he defined it as a true *Puffinus* and a very rare species (*Puffinus gelidus*). This was perplexing, as I have caught any number of them at sea; so I took the bird up to Mr. Salvin, the great Petrel authority, and he at once classed it as above. It is very common off both the Capes, and is easily caught with a roach-hook (or, better still, a small trio hook off a spinning bait) on a light line. From measurements in my book I find the beak is two inches long; wing from anterior bend, twelve inches and a half; length of body, fifteen inches. It is thus summed up by Gould in his 'Birds of Australia':—"Little or no difference is observable in the sexes, but the female is rather smaller than the male; neither did I observe any of the individuals that surrounded the ship to be of a darker colour. In all probability the young attain their normal colouring at the first moult. I quite agree with

II

Capt. Hutton in considering this bird to be allied to the members of the genus *Puffinus.* Crown of the head, ear-coverts, nape and upper surface, tips of the tail-feathers, tips of the under tail-coverts, and the primaries, dark brownish grey; throat, chest, and under surface, white; irides dark brown; culmen and nostrils black; tip of the upper mandible blackish horn-colour; tomia whitish horn-colour; lower part of the under mandible blackish horn-colour; feet white, tinged with blue, the outer toe brownish black." There is a remarkable likeness between this bird and the Greater Shearwater (*Puffinus major*) of our own latitudes. Seebohm tells us a Great Grey Petrel was caught at Swaffham,* in Norfolk, in 1850. Mr. Salvin, on the *Procellariidæ* collected by H.M.S. 'Challenger' (Proc. Zool. Soc. 1878, p. 737), says:—" *Adamastor cinereus* (565 female; 566 male), South Pacific, 5th Nov. 1875. Eyes hazel; feet flesh-colour; the stomach of one was full of the beaks of cuttlefish; stuff from the ship in the other, and small Crustacea." In Lord Lindsay's Expedition Mr. Saunders calls it "Whale-bird." I have often noticed that different ships have different "sailors' names" for sea-birds. In the same way Mr. Saunders calls the Giant Petrel "Cape Hen."

SILVERY-GREY PETREL (*Thalassœca glacialoides*).—This bird, of the genus *Fulmarus*, is almost invariably called at sea the Fulmar Petrel, though as a fact the Fulmar Petrel (*Procellaria glacialis*) does not appear south of the line. Certainly they are much alike, but the bill in my specimen is longer and thinner than that of the Fulmar. Its wings also are longer, and its body lighter. It is, in fact, the southern-seas type of our Fulmar Petrel, and so somewhat differently constituted in order to meet its somewhat different life. They are easily taken with hook and line in the ordinary seafaring fashion. The one now in my collection I caught in lat. 42·48° S., long. 59·43° E., and its dimensions are as follows:— Beak, one inch and seven-eighths long; wing from anterior bend, thirteen inches and a half; length of body, fifteen inches. The general characteristics of the bird are thus defined by Gould :—" All the upper surface and tail delicate silvery grey; outer webs, shafts, a line along the inner webs and the tips of the primaries, and the outer webs of secondaries, slaty-black; face and all the under surface, pure silky-white; irides brownish black; nostrils, culmen, and a portion of the base of the upper mandibles, bluish lead-colour; tips of both mandibles fleshy horn-colour, deepening into black at their points; remainder of the bill pinky flesh-colour; legs and feet grey, washed with pink on the tarsi, and blotched with slaty-black on the joints." In the voyage of H.M.S. 'Challenger' they were met with in the Ice Barrier.

* A marvellous part of the country for rare birds. In 1885 I received from Mr. John Penn, for my collection of British birds, a White-tailed Eagle and a Montagu's Harrier, shot by his keeper close to Swaffham; also seven different species of duck.—J. F. G.

Fulmar Petrel (*Fulmarus glacialis*).—Another *Fulmarus*, and the largest of our British Petrels. It is a rare winter visitant to England, but common in some parts of Scotland. It is well known to all Arctic explorers, and will follow their ship to the highest latitudes, especially if they are whale-fishers. In fact, like *T. glacialoides*, they are particularly at home in the ice, frequenting the Arctic regions, like their cousins do the Antarctic. On the voyage to Australia it would probably only be seen in the English Channel. It is thus described by Yarrell ('British Birds,' vol. iii.):—"In the adult bird the curved point of the bill is yellow, the sides horny white, the superior ridge investing the nostrils greyish-white; irides straw-yellow; the whole head and the neck all round, pure white; the back, all the wing-coverts, secondaries, tertials, upper tail-coverts, and tail-feathers, pearl-grey; wing-primaries slate-grey; breast, belly, and all the under surface of the body, pure white; legs, toes, and their membranes, brownish yellow; the claws slender, but curved and pointed; whole length, nineteen inches; wing from anterior bend, twelve inches; the middle toe and its claw longer than the tarsus."

Spectacled Petrel (*Majaquens æquinoctialis*).—It is a disputed point amongst ornithologists whether this is the same bird as *M. conspicillatus* of Gould. Mr. Salvin, in the notes he kindly gave me, says—"This is the bird of the Cape Seas, and is doubtfully distinct from *M. conspicillatus* of Australia." In the Natural History Museum they are shown as distinct species—*M. conspicillatus* with white spectacles and white throat; and *M. æquinoctialis*, white under the throat only, and called Cape Hen. Mr. Saunders also considers them the same. In his account of the sea-birds collected by Lord Lindsay's Expedition he says:—"The variations in these specimens are rather peculiar. In all the prevailing colour is sooty black; but in the first (spec., No. 56, Sept. 19th, lat. 31° 39′ S., long. 8° 51′ E.) there is a white patch of about three-quarters of an inch in length under lower mandible, and an irregular white streak on the left side, below the line of the gape, but none on the right side; the second (spec. No. 93, Oct. 20th, lat. 32° S.) has rather more white on the throat; and in the third (spec. No. 97, male, Oct. 24th, lat. 29° 45′ S.) the white extends as far back as a line from the eyes." At sea they contest with the Skua Gull the name of Cape Hen, and as their flight, size, and general appearance are much like that of the Skua, perhaps from afar they were originally really mistaken one for the other. The Spectacled Petrel is common down south of the line, especially off such islands as are passed *en route* for Australia. It is thus described by Gould:—"The entire plumage sooty-black, with the exception of the chin, sides of the face, and a broad band which crosses the fore part of the crown, passes down before and beneath, and curves upwards behind the eye, which is white; nostrils and sides of the mandibles, yellowish horn-colour; culmen, tips of both mandibles, and a groove running along the lower mandible, black; irides dark brown."

GREAT WINGED PETREL, ATLANTIC PETREL, SOLANDER'S PETREL (*Pterodroma fuliginosa*).— Mr. Salvin, in the notes he gave me, has marked all these three as *P. fuliginosa*.

Capt. Hutton, writing of the Great Winged Petrel, says:—"This bird, when on the wing, looks very like a huge Swift. It is not by any means common, and I have only seen it east of the Cape of Good Hope. It is not found on Prince Edward Islands nor Kerguelen's Land."—'Ibis,' 1865, p. 286. Gould says it differs from the Atlantic Petrel (*Pterodroma atlantica*) by having longer wings and a greyer face. In the Natural History Museum it is called *Puffinus fuliginosus*, and the long wings are very prominent. Gould names it *Pterodroma macroptera*.

The Atlantic Petrel goes by the name of Cape Parson at sea—probably because of its sombre hue. It is probably *Æstralata carribbœa* of the Natural History Museum. Gould gives its dimensions as follows :—"Total length, fifteen inches and a quarter; bill, one inch and five-eighths; wing, eleven inches and a half; tail, five inches; tarsi, two inches and five-eighths; middle toe and nail, two inches and seven-eighths." He calls it *P. atlantica*.

Solander's Petrel is thus described by Gould:—"Head, back of neck, shoulders, primaries, and tail, dark brown; back, wing-coverts, and upper tail-coverts, slate-grey, each feather margined with dark brown; face and all the under surface, brown, washed with grey on the abdomen; bill, tarsi, toes and membranes, black. Total length, sixteen inches; bill, one inch and three-quarters; wing, twelve inches; tail, five inches and a half; tarsi, three-quarters of an inch; middle toe and nail, two inches and three-eighths." He calls it *P. solandri*.

WHITE-HEADED PETREL (*Æstralata leucocephala*).—*Æstralata* is from *Æstrus*, a gadfly, as applied to the restless flight of this family. They all have the tarsi more or less flesh-coloured. This bird may be recognised at sea by the sort of white patch the head makes against its black wings. It is a marvellous flyer by reason of its extraordinary long and arched wings. It is thus described by Gould:—"Forehead, face, all the under surface, and tail, white; hinder part of the head, back of the neck, and upper tail-coverts, grey; back greyish brown; wings blackish brown; round and before the eye a mark of black; bill and irides black; tarsi, and half the toes and webs, flesh-white; the tips of the toes and webs, black." There is a fine specimen in the Natural History Museum, presented by Sir George Grey, where it is called *Æstralata lessoni*. It is a very powerful-looking bird, with a particularly sharp, curved beak, and looks as if it had been fighting and received a pair of black eyes. In the voyage of H.M.S. 'Challenger,' specimens were obtained in Betsy Cove, Kerguelen; and their black eyes were especially remarked on.

SOFT-PLUMAGED PETREL (*Æstralata mollis*).—Thus described by Gould :—"The sexes are similar in colour, but the young differ from the adult in having all the under surface dark grey, and the throat speckled with grey. Crown of the head and all the upper surface, slate-grey; the feathers of the forehead margined with white; wings dark brown; before and beneath the eye a mark of brownish black; face, throat, and all the under surface, pure white, interrupted by the slate-grey of the upper surface advancing upon the side of the chest, and forming a faint band across the breast; centre tail-feathers dark grey; outer feathers greyish white, freckled with dark grey; bill black; tarsi, base of the toes, and basal half of the interdigital membrane, pale fleshy white, the remainder black. Total length, thirteen inches and a half; bill, one inch and one-eighth; wing, nine inches and three-quarters; tail, five inches; tarsi, one inch and five-eighths; middle toe and nail, seven-eighths of an inch." This is probably *Procellaria roulensis* of the Natural History Museum. A specimen was obtained by H.M.S. 'Challenger' in Nightingale Island:—"Eyes hazel; light bird; 17th Oct., 1875." In Lord Lindsay's Expedition, at the Island of Trinadad they knocked them down with sticks.

WHITE-WINGED PETREL (*Æstrelata leucoptera*).—This bird is thus described by Gould:— "The sexes do not differ in external appearance. Crown of the head, all the upper surface, and wings, dark slaty-black; tail slate-grey; greater wing-coverts slightly fringed with white; face, throat, all the under surface, the base of the inner webs of the primaries and secondaries, and a line along the inner edge of the shoulder, pure white; bill black; tarsus and basal half of the interdigital membrane, fleshy-white; remainder of the toes and interdigital membrane, black. Total length, thirteen inches; bill, one inch and five lines; wing, eight inches and a half; tail, four inches; tarsi, one inch and one-eighth; middle toe and nail, one inch and three-eighths."

COOK'S PETREL (*Æstrelata Cooki*).—Thus described by Mr. Gray, in Gould's 'Birds of Australia':—"Grey above, with the apex of each feather narrowly margined, as well as their bases, white; oblong spot below each eye, wing-coverts, secondaries, and quills, brownish black, with the basal portion of the inner webs of the two last white; the front cheeks, under wing-coverts, and the whole of the under part, white; bill black; tarsi and knee brownish yellow; feet black, with the intermediate webs yellow. Total length, twelve inches and a half; bill, length one inch and seven lines, depth in middle three lines and a half; wings, two inches and a quarter; tarsi, one inch and two lines." A specimen in the Natural History Museum, called *Procellaria Cooki*, is well shown with its wings partly extended.

BLUE PETREL (*Halobæna cærulea*). — Gould says,—"This bird may be distinguished from every other of the smaller Petrels by the conspicuous white tips of the centre tail-feathers." He thus describes it:—"Forehead, lores, cheeks, throat, centre of the chest, and all the under surface, white; narrow space beneath the eye, shoulders, and the outer webs of the first primaries, deep brownish black; back of the neck, side of the chest, back, wings, and tail, grey; the secondaries, scapularies, and six middle tail-feathers, tipped with white; the two outer tail-feathers almost wholly white, and the shafts of all black; bill dull blackish brown, with a stripe of blue-grey along the lower part of the under mandible; tarsi and toes, delicate blue; interdigital membrane flesh-white, traversed by red veins." In the Natural History Museum it is called the Square-tailed Blue Petrel (*Halobæna cærulea*). It is constantly confused at sea with *Prion inctur*, and consequently generally misnamed "Whale-bird."

WEDGE-TAILED PETREL (*Thiellus sphenurus*). In the Natural History Museum this bird is represented by a specimen marked *P. chlororhynchus*. It is thus described by Gould:— "All the upper surface dark chocolate-brown, which gradually deepens into black on the primaries and tail; feathers of the scapularies, which are very broad in form, washed with lighter brown at their tips; face and throat, dark brownish grey; the remainder of the under surface greyish brown; bill reddish fleshy-brown, darker on the culmen and tip; legs and feet, yellowish flesh-colour. Total length, fifteen inches and a half; bill, one inch and five-eighths; wing, eleven inches and a half; tail, six inches; tarsi, one inch and seven-eighths; middle toe and nail, two inches and three-eighths."

CAPE PETREL (*Daption capensis*).—This charming little Petrel abounds in all the temperate latitudes of the Southern Seas, and is universally known by the name of Cape Pigeon. The first I ever saw was very many years ago, and as far north as lat. 11·22° S. For days he followed us, defying our efforts to capture him. When on the feed though (as for instance, after a "Cape blow," when all sea-birds are ravenous), they can be caught in any number; and I remember once a sailor catching four in his hand, so tame and venturesome had they become. I think it is Lord Pembroke, in the 'Earl and the Doctor,' that describes Cape Pigeon-fishing by a sort of rule-of-three sum. As Cape Pigeon-fishing is to Albatross-fishing, so is Trout-fishing to Salmon-fishing—a red quill on a hair-line, instead of a jock-scot on treble-gut, but the *modus operandi* the same. I always found the very best hooks you can have for all these smaller Petrels are the trio, and the smallest you can get; so I strongly advise those about to take a voyage (and interested in this kind of sport) to take plenty of them, from the largest size down to the very smallest. Of course you must cover each of the three barbs with bait, as in Albatross-fishing. Dr. Coppinger was much surprised to see them "dive bodily down, apparently without the least inconvenience," trying to get some submerged morsels of food. I have often had them take a bait like this, and get hooked too. He also says, "A freshly-caught Cape Pigeon, placed on its legs on the deck, seems to forget utterly that it possesses the power of flight, and does not even attempt to use its wings, but waddles about like an old farmyard duck." Like the Albatross, these birds when once attached to a ship seldom leave, if unmolested, until land, or some other vessel, attracts them. I remember a red one (probably caught, painted, and let fly again by some wag) following us for days and days together, and then deliberately deserting us for a homeward-bounder. We were nearing port, and the new ship was leaving it; hence the desertion. The wing makes a very pretty hat-feather for a young lady, which, I am told, often acts as a powerful incentive to the Cape Pigeon fishermen. The bird is thus described by Gould :—" Head, chin, back and sides of neck, upper part of the back, lesser wing-coverts, edge of the under surface of the wing, and the primaries, sooty-brown; wing-coverts, back, and upper tail-coverts, white, each feather tipped with sooty-brown; basal half of the tail white; apical half sooty-brown; under surface white; the under tail-coverts tipped with sooty-brown; beneath the eye a small streak of white; bill blackish brown; irides and feet very dark brown." There are three specimens in the Natural History Museum, showing well the great variation in the markings of the plumage. It is there named the *Pintado* Petrel. Seebohm, in his new ' British Birds,' says a Cape Pigeon was shot near Dublin in 1881. In the collection of H.M.S. 'Challenger' one was obtained in the Ice Barrier, 14th January, 1874.

DOVE-LIKE PRION (*Prion turtur*, Gould).—All the Prions are remarkable for their broad bills, which is especially noticeable in the males. This beautiful little bird is well known at sea as the "Whale-bird," and is so called from its curious broad laminated bill, furnished, like the Right Whale, for feeding on the tiny *Medusæ*. I remember at sea we used to look out for them about the neighbourhood of Tristan-da-Cunha, where they generally appear in flocks. Sometimes, however, they go singly, thereby differing from the Fairy Prion or Ice-bird. I once caught one off the Cape that got entangled in a Mother-Carey line. It is thus described by Gould :—" All the upper surface delicate blue-grey; the edge of the shoulders, the scapularies, outer margins of the external primaries, and the tips of the middle tail-feather, black; small spot before the eye and a stripe beneath, black; lores, line over, beneath and behind the eye, and all the under surface, white, stained with blue on the flanks and under tail-coverts; bill light blue, deepening into black on the sides of the nostrils and at the tip, and with a black line along the side of the under mandible; irides very dark brown; feet beautiful light blue." Mr. Salvin calls both this bird and the Fairy Prion *Prion desolatus*.

FAIRY PRION (*Prion Ariel*, Gould).—This *Prion* is much smaller than the last, and goes at sea by the name of Ice-bird, or rather, I might say, Ice-birds, as I never remember seeing one by itself; they are always in flocks, sometimes of enormous magnitude. It is a pretty sight to see them fluttering up and down like a cloud of silver butterflies, and glittering like the silver tree when its leaves are shaken by the wind. On the voyage to Australia you should first catch sight of them about 40° S. and 10° E., in the region of Tristran-da-Cunha. It almost exactly resembles a small *P. turtur*, only has rather more white about the face, and is generally of a lighter colour. Gould gives the following dimensions:—"Total length, nine inches; bill, one inch and one-sixteenth; wing, six inches and three-quarters; tail, three inches and three-eighths; tarsi, one inch and one-eighth." In the Natural History Museum it is called the Brown-banded Blue Petrel (*Prion desolatus*).

BANKS' PRION (*Prion Banksii*).—Gould describes the bird as very similar to *P. turtur*, but of a longer and more elegant form. Mr. Salvin calls both this and the Broad-billed Prion *Prion vittatus*.

BROAD-BILLED PRION (*Prion vittatus*).—Gould says this bird is rather larger than the last species, and its bill still more dilated. He thus describes it:—"All the upper surface delicate blue-grey; the edge of the shoulder, the scapularies, outer primaries, and tips of the middle tail-feathers, black; space surrounding the eyes and the ear-coverts black; lores, line over the eye, and all the under surface, white, stained with blue on the flanks and under tail-coverts; bill light blue, deepening into black on the sides of the nostrils and at the tip, and with a black line along the side of the under mandibles; irides very dark brown; feet beautiful light blue." There is a good specimen of a male bird in the Natural History Museum which shows the extraordinary breadth of the bill, while a skeleton alongside shows how small the bird is when deprived of its feathers.

SNOWY PETREL (*Procellaria nivea*). — In the Natural History Museum there is a good specimen of this Arctic bird, caught on the Antarctic Expedition, in 77° S. It is pure white, with black bill, black eyes, and yellow legs and feet, about the size of a Cape Pigeon. After a long series of southerly gales it might be met with round Cape Horn. Two females were obtained in the Ice Barrier by H.M.S. 'Challenger' on the 14th of January, 1874.

Concerning the Petrels called *Puffinus*, Gould, in his 'Birds of Australia,' says:—"The flight of the species of *Puffinus* differs considerably from that of the *Procellaria* in being straighter, and performed close above the surface of the water; it is, moreover, so exceedingly rapid that Mr. Davies states it cannot be fairly estimated at less than sixty miles an hour."

Yarrell thus gives their generic characters:—" Bill as long, or longer than the head, slender; upper mandible compressed and curved towards the point; under mandible also slender and curved at the point. Nostrils tubular, opening by two separate orifices. Legs of moderate length, tarsi compressed laterally; toes, three in front, rather long, webbed throughout; hind toe rudimentary. Wings long and pointed, the first quill-feather the longest."

The MANX SHEARWATER (*Puffinus anglorum*).—"This, the common Shearwater that frequents the coasts of Great Britain, is found throughout the North Atlantic Ocean, not ranging into the Baltic, but is found in the Mediterranean as far as the Black Sea" (Dresser). On the Australian voyage we should certainly see one or more on our way down Channel. My specimen is fourteen inches from tip of beak to end of tail; head, back, and wings, black; throat, breast, and under surface, white; beak, legs, and feet, blackish yellow.

ALLIED PETREL (*Puffinus nugax*). — This is the Southern Seas representative of our European Dusky Petrel (*Puffinus obscurus*) of the Shearwater class. Gould thus describes the bird:—"Crown of the head, all the upper surface, wings, and tail, sooty-black; sides of the face, throat, and all the under surface, white; bill dark horn-colour; tarsi and toes, greenish yellow; webs yellowish orange. Total length, eleven inches; bill, two inches and five-eighths; wing, six inches and a half; tail, three inches; tarsi, one inch and a quarter." Both the Natural History Museum and Mr. Salvin call this bird *P. assimilis.*

SHORT-TAILED PETREL (*Nectris brevicaudus*). — This bird is well known at sea as the Mutton-bird. I have often wondered why; but in the 'Field' of January 10th, 1885, I read as follows:—"Mutton-birds (*Puffinus brevicaudus*) are an item of commerce in New Zealand, and caught and potted in their natural oil in immense numbers." Hence, then, I suppose the name. Gould says:—"The sexes are so much alike that they can only be distinguished by dissection. The whole of the plumage sooty-brown; the under surface much paler than the upper; bill blackish-brown, tinged with olive; the under mandible with a longitudinal mark of vinous grey; irides brownish black; outer side of the tarsi and outer toe, brownish black; inner side of the tarsi and two inner toes, vinous grey; webs yellowish flesh-colour, becoming blackish brown towards the extremity." In the Natural History Museum this bird is called *Puffinus brevicaudata.*

FLESHY-FOOTED PETREL (*Nectris carneipes*).—Another *Puffinus*, thus described by Gould:— "There is no difference in the colouring of the sexes, which may be thus described: The whole of the plumage chocolate-black; bill fleshy-white; the culmen and tips of the

K

mandibles, brown; legs, feet, and membranes, yellowish flesh-colour. Total length, fifteen inches; bill, one inch and three-quarters; wing, twelve inches; tail, two inches; middle toe and nail, two inches and a quarter."

In Cassell's 'Natural History' the Petrels are divided into three classes :—The Albatross (*Diomedea*), the largest of all the family; the true Petrels, with long wings and a hind toe always present, birds of sustained flight who swim and dive very little; and the Diving Petrels (*Pelecanoides*), which have short wings and no hind toe. We have already dealt with the first two of these classes, and of the third class there is only one species met with on an Australian voyage, whose great habitat is Kerguelen Island. In the collection of H.M.S. 'Challenger' *P. garnoti* is considered a separate species, but in any case they so resemble one another that the following description will suffice :—

DIVING PETREL (*Pelecanoides urinatrix*).—I first fell in with these wonderful little divers off the Cape of Good Hope, when on board a homeward-bounder. We were cutting the corner extra fine, or should not have sighted them, as they never go far from shore. Their peculiar cry was very noticeable. In Cassell's 'Natural History,' vol. ii., p. 208, the Rev. A. E. Eaton gives the following account of them in Kerguelen Island :—" This bird, as Prof. Wyville Thompson well observes, has a close general likeness to the Little Auk or Rotche (*Mergulus alle*) of the Northern Seas. Both of them have a hurried flight; both of them, while flying, dive into the sea without any interruption in the action of their wings, and also emerge from beneath the surface flying; and they both of them swim with the tail rather deep in the water. But this resemblance does not extend to other particulars of their habits. They had begun to pair when we reached Kerguelen Island. The first egg was found on the 31st of October. Their burrows are about as small in diameter as the holes of Bank Martins (*Cotyle riparia*) or Kingfishers (*Alcedo hispida*); they are made in dry banks and slopes where the ground is easily penetrable, and terminate in an enlarged chamber, on whose floor the egg is deposited. There is no specially-constructed nest. Before the egg is laid both of the parents may be found in the nest-chamber, and very often be heard moaning in the daytime; but when the females begin to sit their call is seldom heard, excepting at night, when the male in his flight to and from the hole and his mate on her nest make a considerable noise. The call resembles the syllable "oo," pronounced with the mouth closed, while a chromatic scale is being made from E to C in the tenor. This kind of Petrel has much difficulty in taking flight from ground which is comparatively level; it is only by running against the wind or by starting from a lump of *Azorella* that the birds are able to rise upon the wing if they happen to alight upon a flat. During my walks on calm nights I used frequently to hear them fluttering along the ground in the dark, and (if I had a lantern) easily caught them by uncovering the light and turning it on them. They flew to light upon H.M.S. 'Supply' on dark nights in October, when there was snow upon the deck." At the Natural History Museum in South Kensington there is a small pamphlet to be obtained, called 'Guide to the · Index Museum Aves (Birds),' by Sir Richard Owen. In page 2, describing the anatomy of birds, he writes:—"The fore pairs of limbs are constructed for the act of flight, and beat as efficiently the denser element, in the few kinds of birds in which those limbs are limited to the act of diving; in both they present the form of 'wings.'" This exactly describes the wings of the Diving Petrel. Dr. Coppinger, in describing the Falkland Island species, says:—"The bill is particularly broad, and of a dark horn-colour; the breast and belly of a dull grey, and the upper parts black; the tarsi and feet lavender. The body is short and plump, and is provided with disproportionately short wings." Speaking of this bird, Darwin says:—"It offers an example of those extraordinary cases of a bird evidently belonging to one well-marked family, yet both in its habits and its structure allied to a very distant tribe." Gould says, in his 'Birds of Australia,' that "it possesses none of those vast powers of flight common to the rest of the Petrels, but has the loss amply compensated for by its powers of diving, which are so great that it is even said to fly under water." He calls it *Pujinuria urinatrix*, and thus describes it:— "Head, all the upper surface, wings, and tail, shining black; ear-coverts, sides of the neck, and flanks, dark grey; all the under surface white; irides very dark greyish brown; base of the cutting-edge of the upper mandible and a line along the lower edge of the under mandible, blue-grey; tarsi and toes beautiful light blue; webs transparent bluish white, tinged with brown; naked pouch hanging from the chin nearly black, and, being very thin, lies in folds like a bat's wing." This little Petrel completes the *Procellariidæ* that we should be likely to meet with on an Australian voyage; and as the Southern Ocean is *facile princeps* their great head-quarters, the list includes a large proportion of the family.

PART II.

PELICANIDÆ.

CHAPTER I.

THE FRIGATE-BIRDS.

HE PELICANIDÆ consist of the Pelicans, Cormorants, Darters, Frigate-birds, Gannets, and Tropic-birds. Out of this remarkably mixed family, only the three last are true Ocean Birds. Taking them in their order, first come the Frigate-birds, of which genus there are two species.

THE GREAT FRIGATE-BIRD (*Tachypetes aquilus*).—It is a common mistake to suppose that "*aquilus*" here refers to any similarity that may exist between this bird and the Eagle. The word is the adjective *aquilus*, signifying dark-coloured, and is used to denote the dusky plumage of the male. *Tachypetes* (from ταχυς, swift, and πετομαι, to fly) means swift-flying. This marvellous swift-flying power of the Frigate-bird is the natural result of a tiny body propelled by wings proportionately larger than any other known bird. Waterton tells us that the muscles of the breast that work these wonderful wings are in themselves one quarter the weight of the whole body. Relying on this power, to get home, it is often met with 1000 miles from shore. Besides its various local names, this magnificent tropical bird has many English aliases; such as "Man-of-War Bird," "Man-of-War Hawk," "Frigate Pelican," and "Great Frigate-bird."

To mariners the Frigate-bird is especially well known from the fact of its being always more or less attracted by any passing vessel. As a rule, the chance of obtaining one of these ocean specimens is small, as they generally fly so high that it would be difficult even to hit them; much more to shoot them so as to fall on deck, which is a feat like dropping a rocketting Pheasant in a space the width of a ship's deck. Once, however, on board a fine sailing clipper, in lat. 12° 13′ N. and long. 111° 3′ E., I was fortunate enough to succeed. Our chief officer had a pointer as shipmate, and coming on deck one night I found the dog rigidly pointing at something apparently high aloft; but as this same attitude was daily taken up for a fly, a pig, or even a swinging block, it by no means followed that anything out of the common was about. On this occasion the object of its attentions was a Frigate-bird, poised on a level with the main truck, and looking for all the world like some huge vampire-bat hovering over the sleeping crew. An easy shot, but dangerous for the ship's braces. Luckily the wire cartridge went like a bullet, and the bird with a broken wing cannoned against the mizen top-gallant sail and fell on the main deck. It turned out to be a mature female, with the reddish-brown throat, white breast, and light brown-speckled edging along the upper wing-coverts peculiar to the sex. From tip to tip of wing it measured six feet one inch. Dr. Bennet says he has frequently seen a Frigate-bird swoop down on a vane at the mast-head and carry it off; so possibly this bird was meditating some such outrage on the ship's property.

With their small and partially webbed feet they are, as may be supposed, but poor swimmers, and, as their principal food is fish, they subsist to a great extent by plundering birds more skilful than themselves in the art of fishing. The principal victim to this aggressive policy is the Gannet, whose dexterously-earned day's sport I have frequently seen appropriated by this handsome buccaneer. Ever on the look-out, wheeling and circling high aloft, the Frigate-bird no sooner beholds the successful headlong dive of the Gannet than he swoops down in hot pursuit to gain possession of the prize. Generally the fish is meekly dropped, and being caught in mid-air by the pirate is carried triumphantly away; but sometimes the Gannet resists, and in such cases always comes off victorious. For though, with its unrivalled flying powers, the Frigate-bird is a masterpiece in the air, the long delicately curved beak and the flimsy claws are of no real use in actual warfare. But bounce and swagger, coupled with a somewhat formidable appearance, in this case as in many another, create a great impression, and birds that could easily hold their own often submit to be robbed rather than fight. Davenport Adams mentions a case where a Frigate-bird was pursuing a small Tern to take possession of his day's fishing, when a larger bird of a different species interfered and drove away the pirate, and then flew away on its own affairs, as if conscious of having performed a good deed. Thus it would seem that the bird-world resent this unnatural behaviour of the Frigate-bird.

Buffon and many old authors depict the Frigate-bird with fully-webbed feet, so perhaps originally it was furnished to fish for its own livelihood; but like the Blind Crabs dwelling in the Caves of Kentucky, that have lost their eyes by disuse, so this bird may now, from the same reason, have lost the means of gaining an honest living. That the disastrous effects of disuse are hereditary there is no doubt. As examples, take the reduced length of wing of the domestic as compared with the wild duck, or the inferior hearing of the tame as compared with the wild rabbit; and, strangest of all, the fact mentioned by Mr. Herbert Spencer, who says that if a silkworm is placed on a mulberry tree it commits the fatal mistake of eating the base of the leaf, thereby cutting through the stem and precipitating itself, leaf and all, on to the ground, and, when there, is unable to remount the tree. It would appear that the Frigate-bird is a mild example of this theory, for certain it is that, though constituted to live on a fish diet, it has now no more chance of catching fish in the ordinary bright still waters of the Tropics than would a bungling fly-fisher on an English trout-stream in similar weather. That it has not the powers necessary for a pirate is equally clear, for if even the small but sharp-billed Tern determined to fight it would defeat the doughtiest Frigate-bird that ever existed. A pirate's life is proverbially a hard one, but terribly so must it be when the pirate is unfitted for the task. Summing up this well-known theory, Mr. Herbert Spencer says, "The dwindling of a little-exercised part has, by inheritance, been more and more marked in successive generations." We may therefore assume that the Frigate-bird will eventually lose its natatory powers. But it is equally true that the effects of use are hereditary, so when this bird has reached the former stage, it is to be hoped, for its own sake, that a stouter bill and more efficient claws will make their appearance.

The Frigate-bird would, then, appear to be in that dangerous transition stage that before now has resulted in the extinction of the genus. This, however, seldom happens unless accompanied by some sudden change in the surroundings. One of the latest instances is the Dodo. Absolutely without foes, it, from disuse, lost all powers of flight. Suddenly its habitations were inundated by man, and, without any time given to establish some other method of escaping, it was annihilated.

I have seen it stated that the Frigate-bird is able to catch the flying-fish as they fly from their numerous piscine pursuers. If this is a fact, there is still a grand legitimate field open to them. On a cloudy, windy day, with the fish well on the feed, Frigate-birds may often be seen fishing for themselves. In such weather, from on board ship, I once watched three of the tribe very successful. Albicore and Bonita were rising freely at the flying-fish scattered by the bows of the moving vessel. So, after much manual labour, I managed to climb along the bowsprit to the outer side of the flying-jib stay. Armed with a patent sea-fly of my own, I hooked an albicore of 36 lbs., that was carefully landed in a potato-sack by a sailor posted below on the dolphin-striker. About to recommence operations, I found myself joined by three gigantic Frigate-birds, looking so fierce and so very near that I felt quite uncomfortable in my rather rickerty position. But I was to be spared as fish were their game, and each soon carried off a struggling beauty in fine style. But on such a day anybody or anything could catch fish, which probably accounted for my own success. Still, while on the subject of this sort of fishing, I would say that I have always found these sea-fish takeable if you have a gut trace and plenty of red in your bait, when they would not look at the ordinary ship's fishing-gear. Most ocean fishing-tackle is too coarse, notably shark-hooks and harpoons. Bring your own on board, of thin sharp steel, also plenty of hooks and lines of all sorts and sizes, and if fond of fishing you will have many a good day's sport.

In the 'Proceedings of the Zoological Society' for 1864 there is a very graphic account of *Tachypetes Palmerstoni*, which is only another name for *Tachypetes aquilus*. It is written by Mr. Wodehouse, from Raiatea, September 3rd, 1863 :—

"'Otaha,' or Man-of-War Hawk (*Tachypetes Palmerstoni*), so called, you know, from its swift and dashing habits. The Otaha does not alight on the surface of the sea, being neither able to swim nor dive; but it hovers over the ocean with unwearied assiduity. Sailors believe it sleeps on the wing. Their flight is easy and graceful, and has the charm of variety. Sometimes the bird may be seen balanced in mid-air, its wings spread apparently motionless, its long forked tail expanding and closing with a quick alternate action, and its head inquisitively turned from side to side to inspect the ocean beneath ; sometimes it wheels rapidly, or darts to the surface of the water, in pursuit of its prey, and at others soars to such a great height that it is lost to sight among the clouds of heaven. When the ocean is turbulent they fare well; but when calm they live by plundering other birds, whose ocean-food they compel them to disgorge by repeated blows, and, when ejected, the Otaha seizes it with great dexterity before it falls into the sea. They are very numerous in these islands. The Otaha builds its nest on the motus, or verdant islets near the reef, amongst the leaves of the 'wild palm.' I believe the female lays no more than three eggs."

In describing the genus *Tachypetes*, Gould says :—"No birds differ more than the members of this genus, for some examples have white and others brown heads, and moreover exhibit many other conflicting differences, both in colour and size."

Mr. Osbert Salvin was kind enough to fully enter into this matter, and explain these differences by showing me his own specimens.

The male bird is all dark blackish-brown (from which the adjective *aquilus*), with a scarlet gular pouch. It seldom exceeds seven feet from tip to tip of wing, and certainly never approaches the fourteen feet one so often reads about.

The female has a reddish brown throat and no pouch, a white breast, and light brown-speckled edging along the upper wing-coverts, supposed to resemble the ports of a ship, and from which the name "Frigate-bird"; all the rest blackish-brown. It is rather smaller than the male. The legs in both are short, thin, and feathered to the toes, which are very slightly webbed, and armed with ridiculously small claws. The thin curved albatross-like bill and long spreading forked tail are also common to both sexes. The rest of the female's plumage is a dusky-brown. The young immature bird of both sexes has the head white, and the baby is an utterly shapeless mass of yellowish down.

Excellent specimens of all these did Mr. Salvin show me. The first three are in the Natural History Museum under the name of *Fregeta aquilus*.

Wilson, in 'American Ornithology,' says the Frigate-bird is not uncommon during summer on the coast of the United States as far south as Carolina.

In the 'Proceedings of the Zoological Society' for 1880 (p. 163), Mr. Saunders describes two specimens obtained in Lord Lindsay's expedition : — "Both females in immature plumage, passing into the adult stage. As this plumage is rarely met with and is little known, it is as well to describe it. The wings, back, and tail are black. with a bar of old brown light-edged feathers along the upper wing-coverts; belly white; flanks and under wing-coverts black; shoulders rusty black, passing into chesnut, which pervades the throat; neck, nape, and crown of head white, slightly tinged with rust; bill horn-white." He also writes concerning this bird, on the same expedition :—" Off Island of Trinidad, South Atlantic, Aug. 20, lat. 20° 23′ S., long. 29° 43′ W. Temperature of air 77° Fahr. and of water 71°. Large numbers seen; some deep black, with scarlet pouch under the throat. Found them sitting on the island."

The black birds with scarlet pouch were, of course, the males. Mr. O. Salvin says it swarms in the Bay of Panama, and breeds in vast numbers on Pajaros Island, in the Gulf of Fonseca.

SMALL FRIGATE-BIRD (*Tachypetes minor*).—I never remember seeing this smaller Frigate-bird on the voyage to Australia, but it is very common all along the northern shores of that great island. In shape and colouring both sexes resemble *Tachypetes aquilus*, only on a somewhat smaller scale. Its habits are also identical. Gould, in 'Birds of Australia' (vol. ii. p. 499), has the following :—

"*Attagen ariel* (Gould, 'Birds of Australia,' vol. vii. pl. 72).—This species of *Tachypetes*, which is considered to be the old *Pelecanus minor* of Gmelin, is rather abundantly dispersed over the seas washing the shores of the tropical parts of the Australian continent, particularly those of Torres Straits.

"The late Commander Ince, R.N., who during the surveying voyage of H.M.S. 'Fly,' was for some time stationed on Raine's Islet, superintending the erection of a beacon, informed me that on his landing on this small island, which is situate in lat. 12° S., at about seventy miles from the north-eastern coast of the Australian continent, and surrounded by a part of the great barrier reef. he 'found this bird breeding in colonies at its S.W. coast, the nest being composed of a few small sticks collected from the shrubs and herbaceous plants which alone clothe the island, and placed either on the ground or on the plants, a few inches above it. The eggs, which are generally one, but occasionally two in number, are of a pure white, not so chalky in appearance as those of the Gannet, and nearly of the same shape at both ends. Upon one occasion I killed the old birds from a nest that contained a young one; on visiting the spot I found the young bird removed to another nest, the proprietors of which were feeding it as if it had been their own; I am sure of this fact, because there was no other nest near it containing two young birds. Some of the eggs were quite fresh, while others had been so far sat upon that we could not blow them; and many of the young birds must have been hatched some two or three weeks. We regarded these birds as the **Falcons of the sea**, for we repeatedly saw them compel the Terns, Boobies, and Gannets to disgorge their prey, and then adroitly catch it before it fell to the ground or water. We never saw them settle on the water, but constantly soaring round and round, apparently on the watch for what the smaller birds were bringing home. I have found in their pouch young turtles, fish, cuttle-fish, and small crabs.' "

In 'The Field,' 4th July, 1885, Mr. H. O. Forbes, describing bird-life in the Keeling Islands, says:—"The Noddy Tern (*Anous stolidus*) and the Gannet (*Sula piscatrix*) were seen in thousands; and he had many an opportunity of noting how their industrious habits are taken advantage of by the swift-winged Frigate-birds (*Tachypetes minor*), much in the same way that the Brown Skuas pursue and harass the English Gulls."

The bird is thus described by Gould:—"The male has the entire plumage brownish black, the feathers of the head glossed with green, and the lengthened plumes of the back with purple and green reflexions; orbits and gular pouch deep red; bill bluish horn-colour; irides black; feet dark reddish brown. The female is similar to the male, but browner; is destitute of the coloured plumes on the back; has some of the wing-coverts and tertiaries edged with light brown, forming a mark along the wing; a collar at the back of the neck; the breast and upper parts of the flanks white, washed with rufous. A nestling bird in my collection is clothed with white down, except on the back and the scapularies, where the dark brown coloured and perfect feathers have just been assumed."

In the Natural History Museum there are specimens of male, female, and young, the latter having the usual light-coloured head.

CHAPTER II.

THE GANNETS.

" But if we would closely mark,
We should see him not *all* dark;
We should find we must not scorn
The teachings of the idiot-born."

ELIZA COOK.

HERE are seven species of Gannet, one of which is found in Europe. They all breed in colonies, and only lay one egg.

THE BROWN GANNET (*Sula leucogaster*).—To describe the birds of an Australian voyage, and omit the confiding Booby would indeed be remiss. Who has not seen him at dusk flapping round the ship on a tour of inspection, previous to pitching on board for the night. Once settled, the bird is yours if you choose to brave a few ferocious digs and snaps, for it will not attempt to escape.

I remember once seeing a ship's cat stealthily stalk a Booby that had boarded us. All her feline craft being brought into play, she soon pounced upon the bird. But the Booby, wildly slashing around with his powerful bill, disentangled himself and hopped aside. And of course flew away, you will say. Nothing of the kind. As far as the bird was concerned, the whole incident appeared to be entirely forgotten. For there it stood, blinking down at the deck, till suddenly aroused by another furious attack from puss. Again, however, did the bird manage to escape, the worse by many feathers; and this went on till—to prevent the death of the Booby—we had to drive away and shut up the cat. Now, although this bird evidently hated the whole performance, it never seemed to comprehend that flying quietly over the side would get over all its difficulties. It was very amusing to see the trouble the cat took over her first stalk, and how soon she found out that in this case all such finesse was entirely thrown away.

It is generally believed on board ship that the Booby sees very badly at night, and is therefore specially glad to come to anchor about dusk. Hence his affection for a ship. This fact is, I believe, generally granted; but it does not account for the stupidity of their behaviour, as they are just as easy to catch in the day-time. It is thus described by Gould, in his 'Birds of Australia,' under the title of *Sula fusca:—*

"The plumage of the two sexes is so precisely similar that it is utterly impossible to distinguish them by external observation; it is true that the colouring of the feet, face, and other soft parts is not always alike, but this difference I believe to be the result of age, rather than of a difference in sex; and if this opinion be correct, the bright yellow-coloured feet are indicative of the bird being fully adult, and the olive-brown of its being immature. Head, neck, breast, all the upper surface, wings and tail dark chocolate-brown; under surface pure white, separated from the brown of the breast by a sharply defined line; irides very pale yellow, blotched before and beneath the eyes with bluish; eyelash light ash-grey; legs and feet pale yellow."

Messrs. Sclater and Salvin, on birds collected by H.M.S. 'Challenger' (Proc. Zool. Soc. 1878, p. 651), describe two specimens as follows:—

"Specimen 141, male, Raine Island. Eyes grey; feet light green; bill bluish towards the base, white at the tip."

"Specimen 510, female, at sea. Eyes white or light grey. Stomach had cuttle-fish; feet yellow with green tinge; bill flesh-colour, cere greenish."

As might be expected of such a bird, it lays its one egg on the bare rocks. Sometimes this dainty is devoured by a Hermit-crab, which allows itself to be sat upon during the process, persuading the boobyish mother it is another egg!

THE RED-LEGGED GANNET (*Sula piscatrix*).—This Gannet may be met with as we near Australia. It is easily recognised by its red legs and the red at the basal part of bill. It is considerably smaller than the rest of the species, and is therefore often called the "Lesser Gannet." It is thus described by Gould, in his 'Birds of Australia':—"The adults have the entire plumage buffy white, with the exception of the wings and tail; the former of which are blackish brown, washed with grey, and the latter pale greyish brown, passing into grey, with white shafts; irides grey; legs and feet vermilion." In the young bird the white is a dull brown.

Messrs. Sclater and Salvin, in the 'Proceedings of the Zoological Society' for 1878, thus describe a specimen collected by H.M.S. 'Challenger':—"No. 123, female, off Cape York at sea. Eyes brown; bill and throat light blue, reddish towards the base, and the tips of both mandibles brown; feet coral-red. The stomach contained cuttlefish about 3½ and 4 inches long."

This bird breeds in *trees.* See 'Ibis,' 1864, p. 379.

THE MASKED GANNET (*Sula personata*).—I have frequently seen these birds in the Southern Seas, and they can always be recognised by a blackish face against a dazzling white body. Together with all the other members of the Gannet family, it possesses the two attributes of magnificent diving at sea and absolute lunacy on board ship. It is thus described by Gould, in his 'Birds of Australia':—"The whole of the plumage of both sexes is pure white, with the exception of the greater wing-coverts, primaries, secondaries, tertiaries, the tips of the two central and the whole of the lateral tail-feathers, which are of a rich chocolate-brown; irides yellow; naked skin of the face and chin in specimen dark bluish black; legs greenish blue."

The correct ornithological name of this Gannet is *Sula cyanops*. Two specimens collected by H.M.S. 'Challenger' are thus described by Messrs. Sclater and Salvin, in the 'Proceedings of the Zoological Society' for 1878:—"137 and 138. Females. Raine Island. Eyes yellow; skin of the throat black; legs and feet slate-colour. Stomach contained fish and cuttlefish."

AUSTRALIAN GANNET (*Sula australis*).—This bird is exactly the same in every particular as our own Gannet (*Sula bassana*). It is another example of a southern type of a northern bird. In this case, however, the resemblance is so complete that two descriptions are unnecessary. Gould gives the following dimensions:—total length, 32 in.; bill, 5½ in.; wing, 19 in.; tail, 10 in.; tarsi, 2 in.

GANNET (*Sula bassana*).—This well-known bird is generally called the "Solan Goose." Describing it, Mr. H. E. Dresser, in his 'Birds of Europe,' says:—"*Sula bassana*, the type of the genus, has the bill longer than the head, straight, elongated, conical, moderately compressed; upper mandible with the ridge broad, separated from the sides by grooves, the sides being slightly convex, with a slender jointed additional piece beneath the eye, the edges sharp, irregularly jagged; tip acute, slightly decurved; the small gular sac is partially bare; nostrils obliterated in the adult, open in the young; wings, long, narrow, pointed, the first quill longest; tail long, wedge-shaped; tarsus very short, sharp behind, scaly; toes united by a membrane, the middle toe longest; claws arched, moderate in size, that on the middle toe pectinate."

It is most interesting to watch a Gannet dive after a fish. They seem to fold their wings, and regularly dash into the sea with a sort of slap-dash that would give a sensitive person a headache. Not content with this, they continue the dive far under the water, regularly pursuing the fish deep down in his own element.

With regard to this wonderful power of descending on a fish, Sir Richard Owen, in his examination of one that died in the Zoological Gardens, says:—"Numerous strips of muscular fibres passed from various parts of the surface of the body, and were firmly attached to the skin; a beautiful fan-shaped muscle was also spread over the external surface of the air-cell anterior to the os furciform. The use of these muscles appeared to be, to produce instantaneous expulsion of the air from the external cells, and by thus increasing the specific gravity of the bird to enable it to descend with the rapidity necessary to the capture of a living prey while swimming near the surface of the water."

The late Frank Buckland, in his 'Notes and Jottings from Animal Life,' also refers to this peculiarity of the Gannet. He says:—"Instead of swooping like a Gull, the Gannet drops almost perpendicularly from a great height into the sea, causing the water to splash up. By an admirable structure the Gannet is enabled to blow the whole of his body full of air, so that, in fact, he becomes an animated balloon, the skin also being divided into air-cells. I am more convinced than ever of the wonderful adaptation of means to ends as seen in the structure of the Gannet."

Taking another extract from this most interesting book, he says:—"Besides the herrings, there were several Gannets caught by the neck in the meshes of the net. The Gannets had dived, as usual, from a great height on to the floating herrings, had thrust their long pointed beaks through the meshes of the nets, and so were drowned."

I take one more extract, this time to show that the Gannet is not altogether such a simpleton as one would suppose:—"I was told quite a new story about Gannets, which breed very abundantly in Scotland. When these birds are building, they steal materials for the nest one from the other. If the thief Gannet is caught in the act, the bird to whom the property belongs gives the thief a good thrashing, which she takes quietly and as a matter of course. If the thief is not detected stealing, she flies out to sea with the stolen property, and then returns looking very innocent, and pretending that she had got it away at sea. So we learn that there are humbugs among birds as among our noble selves."

THE BLACK-TAILED GANNET (*Sula capensis*).—This bird is exceedingly common all round the shores of South Africa, and is consequently well known to Australian voyagers. In Layard and Sharpe's 'Birds of South Africa' we read the following account:—"The 'Malagash,' as it is called by the colonists, or Common Gannet of South Africa. General colour throughout white; the larger feathers of the wings and tail black-brown; the shafts of the former grey; those of the latter white; head and neck, and particularly the back of the latter, ochreous-yellow; space round and before the eye, bare, and of a dark blue colour; a bare stripe of the same extends from the angle of the mouth, on each side of the head, and from the chin, two-thirds of the way down the neck; irides pale fulvous; legs dark livid colour. Length 36″; wing 19″; tail 10″."

Sula variegata. — Concerning this bird, of which very little is known, and which I mention merely to complete the family of Gannets, Mr. Salvin, in a letter to me, says :— "*Sula variegata* is only found along a limited portion of the west coast of South America, but is, I believe, common on the Chincha Islands, being one of the guano producers. I have no doubt whatever it is a very distinct species. It is not at all likely that it would be seen during a voyage to Australia."

CHAPTER III.

THE TROPIC-BIRDS.

" Spirit of ethereal birth!
Aërial visitant of earth!
Take me o'er the proud, blue sea,
Show its beauties all to me."

ELIZA COOK.

IN Mythology, *Phaëton* was the child of the sun; in Ornithology it is essentially the bird of the sun. The hotter the weather, and the fiercer the glare, the more do they appear in their element.

There are three distinct birds of this species. They are distributed throughout the tropics. At sea they go by the name of Boatswain- (bosun) birds, on account of their piping cry.

THE RED-TAILED TROPIC-BIRD (*Phaëton phœnicurus*).—The first Red-tailed Tropic-bird I ever saw was from board ship one brilliant morning, somewhere near the Equator, when pyjama-clad, I was admiring, as a novice, the wonderful beauty of the scene. Compare with our English suns that ball of fire, now rising with a halo of glory that is fast setting ablaze the whole of the eastern horizon; look out at the far-spreading glistening sea dotted with tiny Portuguese men-of-war, catching each breath of wind with their rainbow-coloured sails; watch those glittering shoals of flying-fish, skimming the water to escape the golden-finned albicore, and then plunging into the ocean with a header refreshing to behold; now look straight overhead, and behold, standing out in bold relief against the harebell blue of an absolutely cloudless sky, a pure white fairy-bird, with scarlet beak and tail. You need not ask its board-ship name, for is not that shrill piping cry, now answered by its mate, the very counterpart of Frost, the boatswain, piping "all hands." It is, indeed, the Bosun-bird, welcoming us to his beloved tropics. Some years after, in very different surroundings, I obtained a good specimen of this beautiful bird. Certainly it was in the Tropics, for you seldom see them outside its limits; but it was dull, murky weather, and blowing hard. Notwithstanding the inhospitable elements, a Red-tailed Tropic-bird kept racing round our ship, till a long and lucky shot to windward, aided by the friendly gale, landed him on board.

It now adorns my smoking-room in a case of ocean birds; but though wonderful the skill of the taxidermist, and pleasant the old times these stuffed specimens recall, they all lack the unattainable quality of life, without which they no more resemble their former joyous selves than a mummy does King Pharaoh. All Tropic-birds are very careful of their tails, both in roosting and nesting; but the Red-tailed ones, especially, are frequently met with without those two long appendages; they may be moulting; but it is also the fact that in many of their breeding-stations they are caught for the sake of their brilliant feathers alone, and then set at liberty. Especially are they coveted by the native chiefs, who skilfully work them into a bristling crimson coronet. I was once becalmed off the Island of Ticopia, lat. 12° 14' S., long. 168° 53' E., and boarded by a host of these dusky warriors; and with many of them this same coronet constituted their sole attempt at dress. The Red-tailed Tropic-bird is thus described by Gould in 'Birds of Australia,' vol. ii., p. 503:—"The adults have a broad crescent of black before each eye, the upper part of which extends over and behind that organ; centre of the tertiaries and flank-feathers deep black; the whole of the remainder of the plumage silky white, with a rich roseate tinge, especially on the back; shafts of the primaries black from the base to within an inch of their apex; shafts of the lateral tail-feathers black to within half an inch of the tip; two centre tail-feathers white at the base and rich deep red for the remainder of their length, which extends to eighteen inches, their shafts black; irides black; bill vermilion, with a black streak running through the nostrils, and a narrow line of faint blue at the base of both mandibles; tarsi and the base of the toes and webs faint blue; remainder of the toes and webs black. The young birds for the first year are very different from the adults, being of a silky white, without the roseate blush, with the whole of the upper surface broadly barred with black, and with the black of the shafts of the primaries expanded into a spatulate form at the top of the feathers." In page 502, he says there is "a more decided roseate blush upon the plumage of the male, especially on the back; but this varies slightly in intensity in different individuals of the same sex, and fades considerably in a preserved skin." Dr. Walker, in a voyage from Liverpool to Vancouver's Island, 1863, says:— "From 28° S. to 5° S., flocks of Tropic-birds. One shot that fell on deck (*Phaëton phœnicurus*); from tip of beak to tip of longest ordinary tail-feather, 18¼ inches; one red tail-feather, 9 inches; from tip of beak to extremity of commissure, 3·6; 2·3 to extremity of nasal opening." The eggs (two in number) are of a yellowish white, with reddish spots; but it is extraordinary how greatly specimens differ both in form and colour.

THE COMMON TROPIC-BIRD (*Phaëton æthereus*).—This bird is all white, except black bands on the wings, and a little black at the base of the two long tail-feathers; the beak is red. It is the same shape and size as *P. phœnicurus*, and is met with in the same localities.

THE YELLOW-BILLED TROPIC-BIRD (*Phaëton flavirostris*).—This is a much smaller bird than the two preceding, except in the matter of beak, which is yellow. It resembles *P. æthereus* on a smaller scale. The habits of all three species of *Phaëton* are identical. Mr. E. L. Layard, describing some specimens from the Friendly Islands, says :—"It builds in the forests of Samoa, selecting as a site for its nest the fork of a tree or of a huge perpendicular branch, so that it can enter at one side, and avoid bending or damaging its long tail. I used to watch the flocks going to their sleeping-quarters in the woods, passing high overhead ; and we soon observed that great numbers of them were devoid of tails. One I obtained proved to have moulted this appendage." Dr. O. Finsch says their native name in Ponapé is "Taraki." He describes three specimens, as follows :—"Two males and one female ; the sexes exactly alike ; one male is tinged with very delicate rose-colour. The Island of Ponapé is a new locality for this widely-distributed species."

PART III.

LARIDÆ.

P

CHAPTER I.

THE TERNS.

"And wilt thou, little bird, go with us?
Thou mayst stand on the mainmast tall,
For full to sinking is my house
With merry companions all."

LONGFELLOW.

THERE are about fifty members of the subfamily *Sterninæ*, which Mr. Howard Saunders divides into *Hydrochelidon, Sterna, Nænia, Gygis* and *Anous*. *Hydrochelidon* represents the well-known Marsh or River Terns, which I do not consider Ocean Birds, and we should not expect to meet any members of *Nænia* or *Gygis* on an Australian voyage.

We have therefore only to deal with a certain number of the Sea Terns (*Sterna*) and Noddies (*Anous*). The coloration is very perplexing, as they have three different stages of plumage, *viz.*, the immature, the winter, and the adult breeding state. In the last stage, as a rule, they have black heads. They all have fully webbed feet, armed with sharp claws, a more or less long, thin, straight, sharp bill, and a more or less long forked tail. In length they run from about twenty-two inches down to eight inches. The members of the family that frequent Great Britain are only summer visitants.

At sea all the Terns are called "Sea Swallows." They are much appreciated as generally announcing the approach of the Tropics, and taking the place, astern of the vessel, of the *Procellariidæ*.

Sterna.—Generic characters:—"Bill as long, or longer, than the head; nearly straight, compressed, slender, tapering, with the edges sharp, and the end pointed; the mandibles of equal length, the upper one slightly curved towards the point. Nostrils near the middle of the beak, pierced longitudinally, pervious. Legs slender, naked for a short space above the tarsal joint; tarsi short; toes four, the three in front united by intervening membranes deeply concave in front, or semi-palmated; the hind toe free; claws small, curved. Wings long, pointed, the first quill-feather the longest. Tail forked in various degrees."—*Yarrell, Brit. Birds* (vol. iii.).

THE COMMON TERN (*Sterna hirundo*). — This well-known little bird visits our country from May to September. Before leaving us they collect in flocks, for, like their namesake the Swallow, they seem to prefer taking their long journey south in the largest company possible. Anywhere between Gravesend and the Cape of Good Hope, in the vicinity of land, this sociable little bird might come fluttering out to the ship, and, if undisturbed, be only too glad to spend the night on board, as they often get further out to sea than they intend, and being bad swimmers are glad of a rest.

The *hirundo* for this particular bird, and "Sea Swallow" for all the Terns, is owing to the mutual long forked tail and pointed curving wings, rather than to any resemblance in the actual flying itself, which is of a very inferior order to that of its namesake. I have often watched Swallows beat down a street, and while in full flight pick the flies off the windows, and this without appearing to touch the glass—so wonderfully precise can they be. Now the winged movements of the Tern always seem to me to be of an uncertain and "don't care where I go" sort of order—more like a butterfly than a bird. But perhaps, after all, this is only put on to deceive the young and innocent fry off whom they so dearly love to sup. For how soon this airy, flopperty, style, can be changed into a headlong dash into the water. "What are they after?" I once asked my fisherman on Loch Leven, after watching and admiring a pair of Common Terns taking magnificent headers into the Loch. "Young Perch," he replied, "and I should be sorry to feed my cat on the ones they go for and miss." He was fond of his cat was this man. They are sometimes seen high up swooping along at a tremendous pace, apparently bound on most urgent business. They then fly both strong and fast, and with their powerful wings and long tail much resemble a hawk, from which reason they are often by shore-going mariners called "Sea-hawks." At Hickling Broad by the same class of individuals they are called "Daws."

It is difficult to understand what Yarrell, in 'British Birds,' means when he says they never dive. Probably he only refers to the "head-over-heels" kind of arrangement of the domestic Duck and others. But as to the real genuine header, one might safely back a Tern or a Gannet against any other bird that flies. Both of them seem to drop perpendicularly into the water, and to disappear altogether from view until they emerge with the captured fish.

The mouth of the Eden, on the east coast of Scotland, is a great breeding-place for Terns. The Common Tern there appears to lay three distinct varieties of eggs. A very old St. Andrews naturalist first pointed this fact out to me. He showed me some hundreds of Common Terns' eggs that he himself had collected, which he had divided into three groups, calling them the "grass," the "heather," and the "sand." Sure enough the "grass" egg exactly resembled the striped shadows grass would make, the "heather" was spotted and speckled, and the "sand" was one uniform dark muddy sand-colour. Each egg, in fact, possessed a remarkable resemblance in colour to its surroundings. Is it then that the bird, after choosing a position for its nest, has the power of laying eggs of a colour corresponding to the surroundings? I think not. Is it then that the eggs of themselves take up this colour? Certainly not, or all eggs would be similarly affected, and this we know is not the case. No; it rather appears to be another example of Darwin's "survival of the fittest." Taking into consideration the now well-known fact that the dispositions and actions of animals are hereditary, a long ancestral line of birds would year after year build in the same sort of locality, and those eggs resembling that locality would be the most likely to escape destruction from their numerous enemies, and so eventually establish three varieties of the Common Tern laying eggs like their surroundings. One would therefore be inclined to say that such a disposition in time must constitute sufficient specific distinctness in the bird itself to form three separate species. Darwin tells us that the Tiger and Leopard are branches from the same stock. Two varieties were spared : those most resembling the stripes like the shadows of the tall jungle-grass, and those most resembling the spotted appearance of the shadows caused by leaves—eventually forming the two present distinct species, though at what precise moment a variety turns into a species is not quite clear.*

While on the subject of Terns' eggs, I may mention a story a very keen old collector once told me. He had found some Terns' eggs very much "sat on," and taking them home he hatched them off in hot water. He then put the young birds in a basket by the fire, when, to his horror, a cat immediately took possession of the brood—not, however, to eat, but to rear, which she did so successfully that they all grew up and finally flew away out of the window. This sounds like a fairy tale, but such was the story, and he seemed a truthful old man.

The bird is thus described by Yarrell (Brit. Birds, vol. iii. p. 308) :—"In the adult bird in summer the bill is coral-red, the point black, irides dark brown ; back and wings ash-grey; outside web of the first primary slate-grey, the shaft white, inner web light

---

* The *old* rule was that the hybrid offspring of species are sterile, and the mongrel offspring of varieties are fertile.

grey; tail-coverts white; outer webs of tail-feathers pale ash-grey, inner webs white; chin, neck, breast, and under surface dull white; legs, toes, and membranes coral-red. The whole length of the bird fourteen inches and a quarter; from the wrist to the end of the longest quill-feather ten inches and a half. A young bird killed in August has the point of the beak dark brown, the base reddish-yellow; forehead dull white; posterior part of the crown, the ear-coverts, and the occiput black; chin and neck all round white; back and wing-coverts ash-grey, each feather margined with ash-brown and white; outer web of the first quill-feather black; the others ash-grey; under-surface of the body white; legs, toes, and membranes reddish-brown."

Though this bird has been separated from the Arctic Tern since the days of Linnæus, there is no doubt but that the Arctic is really the commonest Tern in our islands. Mr. Howard Saunders and others therefore call it *Sterna fluviatilis*, and drop the *Sterna hirundo*, as that name had become common to both Terns. Dresser says:—"The Common Tern is generally distributed throughout Europe, though hardly so abundant as the Arctic in the northern portion of the Continent. Like many of the *Sternæ*, it is common to both the Old and New Worlds, and even extends into Southern Africa."

THE ARCTIC TERN (*Sterna arctica*).—One thing is certain, that the English name of this bird should have been the Common Tern, if there is to be any meaning in the word "common." Mr. H. Saunders takes the name *Sterna macrura* for this bird, and concludes his article as follows:—"The Arctic Tern ranges along the coasts of Northern Europe, Asia, and America; in winter it visits the African coast, descending as far as Walwich Bay; and I have an example obtained by Wucherer off Bahia, the only instance known of its occurrence so far south on the American side."

It is thus described by Yarrell ('British Birds,' vol. iii. p. 408):—"The adult bird in summer has the bill coral-red, the extreme point sometimes black; forehead, crown, and nape black; back, wings, and wing-coverts pearl-grey; outer web of the first primary lead-grey; tail-coverts and tail-feathers almost white, the two longest tail-feathers on each side grey on the outer webs; cheeks white; chin and upper part of neck in front, and on the sides ash-grey; breast and all the under surface of the body as dark a grey colour as that of the back; legs, toes, and their membranes orange-red. The whole length of the bird from the point of the bill to the end of the middle, or short, tail-feather twelve inches and a half, to the end of the longest tail-feather two inches and a half more, or fifteen inches whole length; from the wrist to the end of the longest quill-feather, eleven inches; length of the tarsus only half an inch."

Mr. H. Saunders, in explaining how perplexing are the variations of the colouring of the Tern family, states as an example that the red bill of both this Tern and the Common Tern becomes quite dark in the first two weeks of October.

The bill of the Arctic Tern is shorter, slighter, and rather more curved than that of the Common Tern. Dresser says this Tern is found as far south as the Cape of Good Hope.

CASPIAN TERN (*Sterna caspia*).—This well-known bird, the largest of the British Terns, is occasionally met with on the east coast of England. Mr. Howard Saunders says it is found from Northern Europe to New Zealand, and in America from Labrador, where it breeds, down to New Jersey. Dresser says:—"Ranging from Northern Finland down to South Africa, and frequenting the coasts of Asia and the Islands to New Zealand, the Caspian Tern is also found in the Nearctic Region from the shores of Labrador down to New Jersey, and has therefore a very extensive range."

In reading over one of my old logs I find mention of a strange bird with a whitish head and long beak that flew on board, and by its actions and appearance completely puzzled all the ornithological talent of our ship. We perceived it was a species of Tern, but we were amazed at its size and perfect self-possession. It allowed itself to be caught and fed, and when released still kept to the ship, much to the disgust of certain dogs and cats, who were not allowed to touch it. Finally it took its departure in that airy and unconcerned way peculiar to the Tern family. At the time it was certainly a case of—

> "Although I do not know your name,
> Nor can I tell from whence you came,"

but afterwards, reading over its description and looking at the sketch I made of it at the time, I came to the conclusion it was a Caspian Tern in winter plumage. It is the only one I have ever seen, and its friendliness and general good faith alone saved it from becoming an addition to my collection of birds.

As this happened close to Australia, it would be the Caspian Tern of Australian waters. Gould gives it then a different name, calling it the "Powerful Tern," *Sterna strenua*, and says it is larger than our bird, and has a more richly coloured bill. Mr. H. Saunders, however, says that it is now generally conceded that there is but one species. Its range must, therefore, be very large, as it is occasionally met with in Great Britain. Owing to its great size, it is sometimes called the "King of the Sea-swallows."

The bird is thus described by Yarrell, 'British Birds' (vol. iii. p. 387):—"When in their summer plumage the bill is vermilion-red, lighter in colour at the points; the irides reddish-brown; forehead, at the top of the head, and the nape of the neck rich black, the feathers of that colour on the occiput elongated; lower part of the neck all round white; the back and all the upper surface of the body, the wings, and tail-feathers ash-grey; the first six wing-primaries of a much darker grey, a slate-grey, with white shafts; the tail but slightly forked; the chin, throat, breast, and all the under surface of the body pure white; legs, toes, their membranes, and the claws black, the latter strong and curved. The whole length of the specimen described, from the point of the beak to the end of the long feathers of the tail, nineteen inches; some specimens measure twenty to twenty-one inches. Adult birds in winter have the head white." Seebohm says the forehead keeps the same dark colour both summer and winter.

The ALLIED TERN (*Sterna affinis*), called by Gould the "Indian Tern," is a big edition of the Sandwich Tern with a yellow bill. Bree says it is the only European Tern that is not a visitant to Great Britain. Mr. Howard Saunders, calling it *S. media*, says:—"It ranges from the Straits of Gibraltar, along the Mediterranean, down the Red Sea to Madagascar; and eastwards along the Indian coast and islands, throughout the Malay Archipelago, the Aru Islands, down to Torres Straits and Port Essington." He says it can always be distinguished by its pearl-grey rump and tail. The bird is thus described by Bree in 'Birds of Europe' (vol. v. p. 56):—"Male and female in breeding plumage have the forehead, vertex, and occiput of a deep black; nape silvery white; top of the body bluish ash, like the Sandwich Tern; lower part of the body, front and sides of the neck, and cheeks of a silvery white; wing-coverts like the back; primaries of a velvety ash, bordered on their inner webs with white; tail bluish ash, darker than the wing-coverts, with the most lateral quill on each side of a velvety ash; beak yellow and slightly 'Gull-billed'; feet black."

The BASS'S STRAITS TERN (*Sterna poliocerca*) is another large yellow-billed Tern, closely allied to the last. Gould gives its total length, 17½ inches; bill, 2¾; wing, 12¾; tail, 7; tarsi, 1. Mr. Howard Saunders, describing it under the name of *S. bergii*, says:—"The distinguishing character of the Large Sea Tern is the white band of feathers across the base of the bill. In the adult plumage, and even in winter plumage, there is no other species of its size in which the mantle and tail are of so dark a grey." On the Australian voyage this bird might be met with anywhere near land south of the line, right round the Cape of Good Hope, to Australia. It is thus described by Gould, 'Birds of Australia' (vol. ii. p. 396):—"Crown of the head and occipital crest jet-black; forehead, back of the neck, and all the under surface silky white; back, wings, and tail grey; secondaries tipped with white; shafts of the wings and tail white; bill yellow; irides black; legs and feet brownish-black." This bird abounds on the shores of Tasmania and New South Wales. In the P. Z. S., 1864, Part I., January to March, is the following description of this bird by Mr. Wodehouse (dated, "Raiatea, 3rd September, 1863"): — "'Otino' (*S. poliocerca*), White Reef-bird, also a species of Heron, as you will have seen. This gentleman passes his time on the 'barrier-reef,' amidst the foam of the broken wave, which brings with it from the ocean the small fish which constitute his food. His home is, too, the 'wild palm' of the 'green motu,' close to his beloved reef, on whose wave-beaten surface he passes his life. I do not know how many eggs the female lays."

The TORRES STRAITS TERN (*Sterna cristata*) is closely allied to the two former. Mr. Howard Saunders describes it under the heading of "*Sterna maxima.*" Comparing it with the Caspian Tern, Seebohm says it is a smaller and distinct species, and may be distinguished by having the outer portion of the inner webs of the first six primaries white. On the Australian voyage it would be seen along the West Coast of Africa, near the line, and on the East Coast of Australia.

The SOUTHERN TERN (*Sterna melanorhyncha*) is common off the shores of New Zealand. It is *S. frontalis* of Mr. Howard Saunders. Space round eye and occiput black; forehead, sides of the neck, and under surface white; upper surface, back, wings, and tail grey; bill black. Gould gives following dimensions:—Total length, 13 inches; bill, $2\frac{1}{4}$; wing, $9\frac{3}{4}$; tail, $6\frac{1}{4}$; tarsi, $\frac{3}{4}$.

Sterna hirundinacea is described by Mr. Howard Saunders "as the largest and lightest in colour of the medium-sized Sea Terns; and the entire bill (which is long and powerful) is bright red in the adult." On the homeward voyage round the Horn it should be looked out for off the Falklands.

Sterna vittata is very similar to above, only smaller. It frequents St. Paul's Island and Kerguelen Island.

Sterna virgata also frequents Kerguelen Island. Mr. Howard Saunders says:—"The bill is rich blood-red, and the feet are red."

Sterna antarctica is thus described by Mr. Howard Saunders:—"This species, of a nearly uniform smoke-grey colour, appears to be confined to New Zealand, and principally to the South Island, where it deposits its eggs on the bare ground, making no nest. The bill and feet are orange."

R

SANDWICH TERN (*Sterna Boysii*).—Another British Tern, and fairly common all round our shores in the autumn. I have frequently seen them along the coast of the Firth of Forth. Dresser says:—"Like most of the sea-birds, the present species of Tern has a very extensive range, being found in Europe, Africa as far south as the Cape of Good Hope, and America as far south as Brazil."

Under the name of *S. cantiaca*, Mr. H. Saunders says that the range of this species is from Northern Europe to the Cape of Good Hope and Bay of Bengal in winter; and along the Atlantic coast of North America to the West Indian Islands, Honduras, and Brazil, as far as Bahia.

It is thus described by Yarrell, 'British Birds' (vol. iii. p. 391):—"The adult bird in summer has the bill black, the tip yellowish white; the irides hazel; all the parts of the head above the eyes black; the feathers on the occiput elongated, forming a loose plume which ends in a point; cheeks, sides, and bottom of the neck behind white; back and wings ash-grey, the ends of the tertials almost white; the longest primary slate-grey, with a strong and broad white shaft; the next two or three primaries each a little lighter in colour than the first, and diminishing in colour in succession till they become of the same tint as the wing-coverts; the tail white and forked; chin, throat, neck in front, and all the under surface of the body pure white; legs, toes, and their membranes black, claws curved and black. The whole length of the bird, from the point of the beak to the end of the longest quill-feather, eleven inches; the first quill-feather the longest in the wing. Adult birds in winter have the head white."

THE ROSEATE TERN (*Sterna dougalli*).—This lovely Tern is fast disappearing from our coasts, and there are only two or three localities where it now breeds; perhaps the principal of them is the Fern Islands, off the coast of Northumberland. Mr. H. Saunders says:— "Apart from its light and elegant shape, and its proportionately short wings, this species may always be recognised by the white inner margins of the primaries, extending quite round the tips of the feathers as far as the outer webs; the rump and tail-feathers are washed with grey. The coloration of the bill varies considerably with age and seasons; in some specimens it is black almost to the base, while in others the red or orange extends far in front of the angle."

Mr. Howard Saunders says the GRACEFUL TERN (*S. gracilis*) of Gould is merely a form of *S. dougalli*, with more red than usual in its bill. Amongst other places, this bird is to be found along the Mediterranean, in Ceylon, off the Cape of Good Hope, and along the West Coast of Australia.

It is thus described by Yarrell, 'British Birds' (vol. iii. p. 395):—"In the adult bird in summer the bill, from the point to the nostrils, is black, from there to the base or gape red; the irides dark; all the top of the head black; neck all round white; back, wing-coverts, and quill-feathers ash-grey; the outer webs of the primaries dark grey, the inner webs lighter; tail-feathers very long, extending beyond the ends of the wings, the colour pale ash-grey; breast and all the under surface of the body white, strongly tinted with a delicate rose-colour, whence the bird derives its name; legs, toes, and their membranes red. The whole length of the bird, fifteen inches and a half. The plumage of the adult bird in winter is unknown; but it is probable that it only loses the black and the rose-colour, which belong to the breeding-season."

LESSER TERN (*Sterna minuta*).—This, the smallest of the Terns that visit our shores, is the type of a group of small Terns, called by some *Sternula*. I have often watched the Lesser Tern in Scotland paddling along on a sandy shore, hunting for the dainties exposed by a falling tide. They lay two or three eggs in any depression in the ground. Those in my collection were taken by the River Eden, N.B., just above high-water mark. They are often met with at sea in the vicinity of land; and I have a specimen in winter plumage that I caught on board ship off the Island of Ascension.

Dresser, in 'Birds of Europe,' says:—"The range of this, the least of our European Terns, is not so extensive as that of some of its allies. It is met with throughout temperate Europe, occurring in winter on the coast of West Africa as far as the Cape of Good Hope; and it is also found in Western Asia."

Mr. H. Saunders says:—"This Tern, which has *dark* shafts to the outer primaries, and the rump and tail *white*, ranges through temperate Europe to India; occurs in winter on coast of West Africa as far as the Cape of Good Hope, whence there is a specimen in the British Museum."

It is thus described by Yarrell, 'British Birds' (vol. iii. p. 412):—"In the adult bird in summer the beak is orange, tipped with black; irides dusky; forehead white, crown of the head and the nape jet-black; back and wings uniform delicate pearl-grey, the first, second, and sometimes the third primary slate-grey; upper tail-coverts and tail-feathers white, tail forked; chin, throat, sides of the neck, breast, and all the under surface of the body pure white; legs, toes, and membranes orange. The whole length of the bird rather more than eight inches; from the wrist to the end of the wing, six inches and three-quarters. The adult bird in winter only varies in having the head dull black, instead of deep black." The young bird is speckled on back and head, with bill and legs pale brown.

The Southern Seas representative of the Lesser Tern is called by Gould, *Sternula nereis*. He says:—"*Sternula nereis* is a beautiful representative in the Southern Ocean of the Lesser Tern of the European seas, the habits, actions, and economy of both being precisely the same." It is, however, rather larger, like most of the southern representatives. Describing it, Mr. H. Saunders says:—"This species, which appears to be confined to Australia and New Zealand, may be distinguished from the other small Terns by its somewhat larger size; the pale grey of the mantle, and especially of the primaries; and by its having *no black lores*, but only a dark spot in front of the eye." The adult bird is ten inches long.

A little Tern, almost exactly similar to our Lesser Tern, is *Sterna antillarum*, concerning which Mr. H. Saunders says:—"Similar to the above (*S. minuta*), and has also *dark* shafts to primaries; but *the rump and tail-coverts are pearl-grey*, like the mantle; and there is but little black at tip of bill. Ranges throughout temperate America, on both coasts, and down to the Antilles, Trinidad, lat. 10° N."

Another of these little Terns is *Sterna sinensis*, thus described by Mr. Howard Saunders:— "Like *S. minuta*, but *shafts* of outer primaries *white*; as a rule also the bird is a trifle larger and stouter, and has a longer development of lateral tail-feathers than *S. minuta*." In the 'Proceedings of the Zoological Society,' 1875, Part III., the bird is thus described:— "With one blackish primary, from 7·25 to 6·9 inches in length; the bill long, and not exceeding 1·3; vent and shorter under tail-coverts light iron-grey; feet clear orange." On an Australian voyage would probably be only seen as we approached Australia.

Sterna balænarum is thus described by Mr. Saunders:—"In this species there is no white frontlet, the black feathers coming down to the base of the bill, which is slender and black, except at the gape; the tail is grey, like the mantle; and the tarsi and feet are the smallest of those of the group. The shafts of the primaries are *white*. Wabirch Bay to the Cape of Good Hope is its range, so far as is known."

These complete the list of little Terns (called sometimes *Sternula*) that we should be likely to meet with on the voyage to Australia. There are several species of this group peculiar to the West Coast of America.

THE GULL-BILLED TERN (*Sterna anglica*).—This Tern may always be recognised by reason of its bill being of the Gull-shape. As its Southern Seas representative (*S. macrotarsa* of Gould) is identically the same, this bird has a most extensive range. Mr. Saunders says:—From Western Europe to the China seas, throughout India, Ceylon, and the Malay region down to Australia, and along the East Coast of America as far as Patagonia; on the Pacific side it has only been observed in Guatemala."

This Tern has occasionally been met with in Great Britain. It appears to have been originally confounded with the Sandwich Tern, but first pronounced a distinct species by Montagu, who named it *S. anglica*, not knowing it existed elsewhere.

It is thus described by Yarrell, 'British Birds' (vol. iii. p. 409):—"In the adult in summer the bill is black, and one inch and a quarter in length from the point to the feathers on the forehead; the angle at the symphisis of the lower mandible rather prominent; irides reddish brown; forehead, crown, and nape jet-black; neck behind greyish white; back, scapulars, wings, the coverts, secondaries and tertials, upper tail-coverts and tail-feathers uniform pale ash-grey; the outside web of the first primary slate-grey, the other primaries pearl-grey; chin, throat, breast, and all the under surface white; legs, toes, membranes, and claws black. The whole length of the bird figured from and described, fifteen inches and a half; wing from the wrist, thirteen inches." In winter the head is white.

THE SOOTY TERN (*Sterna fuliginosa*).—There are several varieties of Sooty Terns, but this bird and *Sterna anæstheta* are the only two we should come across on an Australian voyage. These two are very much alike; but Mr. Howard Saunders explains to us, in his article on the *Sterninæ*, that they may always be recognised, amongst other distinctions, by their different feet. The Sooty Tern (*S. fuliginosa*) is a wonderful flyer, and has about the largest geographical range of all the Terns. It has been shot in England, and is very plentiful round Australia. The Island of Ascension is one of its great head-quarters. There they assemble in thousands, and are called "wide-awakes." Wilson says, in Capt. Cooke's voyage, it has been seen 100 leagues from shore. Thus in many parts of an Australian voyage it is almost certain to be met with.

Gould thus describes it (under the name of *S. serrata*) in 'Birds of Australia' (vol. ii. p. 408):—"This common species appears to be very generally distributed on the seas surrounding Australia, but to be less numerous on the southern than on the western, northern, and eastern coasts. It is now supposed to be the same species which frequents the shores of the countries washed by the Atlantic, both north and south, and that examples from North America and Australia are not different; if this be the case, no bird of its family enjoys so wide a range over the globe. The colouring of the species is as follows:—

s

Lores, crown of the head, and back of the neck deep black; the apical half, the shaft, and the outer web of the lateral tail-feathers white, passing into grey on the lower part of the abdomen and under tail-coverts; irides dark brown; bill black; feet brownish black."

Yarrell tells us one was shot in October, 1852, at Tutbury, Burton-on-Trent, and therefore includes it in his ' British Birds.' He thus describes it :—" The whole length of the bird, fourteen inches and a half; wing, from flexure, eleven inches, and extending one inch beyond the end of the tail, but in the adult bird the outside tail-feathers on each side extend for two inches beyond the ends of the closed wings, giving a length of seventeen inches to the fully-adult bird; the leg and middle toe equal in length, each measuring one inch."

Dresser says :—" Found numerously on the southern coast of the United States and of Central America; the present species is common on some of the islands in the Atlantic, on parts of the African coast, being somewhat rarer on the coast of Asia, though tolerably common and generally distributed in the Australian seas; but to Europe it is an extremely rare straggler."

At the January Meeting, 1886, of the Zoological Society, Mr. Howard Saunders exhibited an adult specimen caught alive, near Bath, October, 1885, and pointed out that only two examples of this species had as yet occurred in Great Britain.

Sterna anæstheta is called by Gould the PANAYAN TERN, *S. panayensis.* He says it is common on the West Coast of Australia. Mr. Howard Saunders gives it the same geographical range as *S. fuliginosa.* It is thus described by Gould, ' Birds of Australia ' (vol. ii. p. 413):—" Forehead, line over the eye, chin, and throat white; lores, crown of the head, and nape black; back, wings, and tail light sooty brown, the outer tail-feathers being white at the base and on the outer web for two-thirds of its length; edge of the shoulder and under surface of the wing white; under surface white, slightly washed with grey; irides blackish brown; bill black; legs and feet blackish green."

Seebohm calls this bird the Smaller Sooty Tern.

Genus ANOUS, *Leach.*

"The Noddies," remarks Mr. Jerdon, "are well-known oceanic birds, frequenting tropical and juxta-tropical seas. They differ from most Terns in their even or somewhat rounded tails; and still more in the manner of their flight, which is steady and slow. They settle on the water when taking their food, which consists chiefly of mollusks and fatty matter; and they are very silent birds. Sundevall, who noted these differences, states that in their mode of life they resemble Petrels rather than Terns."

"Unlike other Terns which frequent the sea-shores and rivers, the Noddies frequent the wide ocean, far remote from land, and which, like the Petrels, they seldom quit, except at the breeding-season, when they congregate in vast multitudes on small islands suited to the purpose."—Gould, 'Birds of Australia,' vol. ii. p. 412.

Audubon says the Noddies never rest on the ground; but there appears to be a difference of opinion on this subject.

Mr. Howard Saunders describes five members of the genus. On the Australian voyage we should only expect to meet two.

NODDY TERN (*Anous stolidus*).—The following interesting account of several ocean birds (including this bird) appeared in the 'Field,' June, 1886, in an account of the Chesterfield Reefs, by Mr. Layard:—

"Long Island is about one and a half miles in length, by one hundred yards across in the broadest place. About a mile at one end is thickly covered with trees, all of an equal height of about forty feet. I do not know what botanical genus they belong to, but the wood was so brittle that a branch as thick as my leg could be snapped in two as easily as breaking a carrot, so that climbing for eggs was a dangerous amusement. When the wood dried, it was so friable that it would not burn. The tree had bunches of seeds so glutinous that I often found the wretched Terns dying of starvation on the ground, with their wings and tails so tightly glued together that they could not fly, and it took me some trouble to release them. It was a curious sight; whenever a nest could be placed on a forked branch or a projecting knot of a tree-trunk, there a nest was. It was a rather thick but perfectly flat structure of dried leaves or seaweed, but round about the houses the birds picked up any old rags, bits of paper, twine, &c. Every nest had a bird, old or young, on it, and the screaming from the millions of Terns was something deafening. The greater number of the birds were the white-headed sooty-bodied Terns (*Anous melanops* and *A. stolidus*); but there were also thousands of Great Boobies (*Sula fusca* and *S. personata*) and Frigate Birds, who built their large nests on the tops of the trees, where it was almost impossible to get at them. The beautiful Tropic Birds, or 'Boatswains,' had all been driven off the island by the Malabar guano-diggers, who killed them on their nests to get their feathers to stuff mattrasses and pillows. The whole island also was completely honeycombed with the burrows of the Black Petrels, or Mutton Birds (*Puffinus brevicaudus*)."

Yarrell tells us that the reason this bird is so often caught on board ship is "because it does not see well by night, and it is for this reason it frequently alights on the spars of vessels, where it sleeps so soundly that the seamen often catch them."

Mr. Howard Saunders, in his article on the *Sterninæ*, says of this bird:—"This well-known species, a straggler to the British seas, ranges from the gulf-coast of North America to the shores of Australia, throughout Polynesia, and occurs, in fact, in all tropical waters. There appears to be no constant difference between individuals from the most distant localities; and this similarity applies to its habits and breeding, its single egg being deposited on a nest of seaweed placed on mangrove bushes, in the fork of a tree, or even on the bare rock."

The bird is thus described by Yarrell, 'British Birds' (vol. iii. p. 420:—"In the adult bird the bill is black, from the base of the bill to the eye is also black; irides brown; the forehead and crown buff-colour; occiput smoke-grey; the whole of the body above and below and all the wing-coverts dark chocolate-brown; primaries and tail-feathers brownish black; legs, toes, membranes, and claws black. The whole length of the specimen here figured and described, fourteen inches and a half to the end of the tail, which is graduated, the middle pair of feathers being the longest; the wing, from the carpal-joint to the end of the first quill-feather, ten inches and a half."

LESSER NODDY (*Anous melanops*).—This Noddy is very abundant in the Australian seas. Gould accounts for this from the fact that they roost up high, and thus escape the attacks of a small lizard that destroys multitudes of the other Noddies. He thus describes it in 'Birds of Australia' (vol. ii. p. 419):—"Crown of the head and back of the neck light ash-colour, passing into deep grey on the mantle and back; immediately before the eyes a large patch, and behind a smaller one, of jet-black; posterior half of the lower and a smaller space on the upper lash snow-white; throat, fore part of the neck, and all the under surface deep sooty-black; wings and all the under surface of the same colour, but rather browner; bill black; tarsi and toes brownish black. Total length, 12 to 13 inches; tail, 2¼; wing, 8¾; tail, 5; tarsi, ⅞; middle toe and nail, 1½."

CHAPTER II.

THE GULLS.

"Away on the winds we plume our wings,
And soar, the freest of all free things;
Oh! the Sea Gull leads a merry life
In the glassy calm or tempest strife."

ELIZA COOK.

R. HOWARD SAUNDERS, in his exhaustive article in the 'Proceedings of the Zoological Society,' 1878, Part I., enumerates forty-nine different species of Gulls, fifteen of which may be called British and three Australian. Though the Gulls are more birds of the sea-shore than the ocean, a ship would certainly be followed by various members of the family all down the Thames, the English Channel, and well into the Bay of Biscay; also on the voyage the ship would be visited, when nearing land, by those Gulls peculiar to the locality; and lastly, the Australian Gulls would accompany the vessel from Cape Otway to its destination.

In the following pages I have endeavoured to describe twenty-two different species that we might thus fall in with on an Australian voyage. This includes all the Gulls of both Great Britain and Australia.

The individual variations in plumage of this subfamily are even more perplexing than the Terns, as, besides the seasonal changes, they are a matter of three years in assuming the complete adult dress.* This is a wonderful provision of Nature to enable them to gain experience in the ways of the world (clad in a less conspicuous garb) before exposing themselves to its dangers in a more marked and easier discerned plumage. But for a similar thoughtful arrangement the Golden Pheasant must have long ago become extinct. The males are considerably larger than the females, but alike in plumage. Some of the smaller Gulls resemble the Terns in assuming a dark-coloured head during the breeding-season. They assemble together in vast numbers for breeding purposes, nesting on the rocks; or inland,

---

* I suppose a Darwinian would say that this is the result of the gradual survival of only those birds that postpone their adult plumage longest.

T

near rivers and lakes, on the ground; and sometimes, but rarely, in trees. They appear to sleep equally well in three positions,—standing on one leg, squatting on the ground, or floating on the water.* Mr. Howard Saunders divides this great subfamily into five distinct genera, of which more anon.

THE COMMON GULL (*Larus canus*, Linn.).—For many years I kept three Common Gulls in a garden, and most entertaining they were. They spent most of their time on a lake, but always came up to the house to be fed, and tapped vigorously at the windows if no food was forthcoming. Some were pinioned, some not. Those that could fly would take long excursions, but always returned at night. They all had a rooted aversion to dogs, especially one that had been retrieved twice one morning by an officiously-clever retriever: clever because, though the Gull was caught and carried about a hundred yards, it was not in the least hurt; officious, because nobody wanted it retrieved at all, as it was quietly walking about in its own arena. It was, however, done so well that when some time afterwards I lost a brood of Golden Pheasants, I employed the same dog; and he found the whole number, one by one, squatting in different parts of the garden, and carried them back to their foster-mother, unhurt. Unfortunately for the Gulls, our chickens began mysteriously to disappear. Two, of a different species (Lesser Black-backs), were at last caught red-handed, striding off at a tremendous pace with a chicken apiece; and the edict went forth to get rid of all the Gulls. So perhaps the innocent suffered with the guilty. At the sea-side their favourite spot seems to be the mouth of a sewer; in shore they love following the plough to feed on the newly-turned-up delicacies; our specimens simply ate anything that was given them; so with such an "all-round" appetite they may very likely have occasionally grabbed a chicken on the sly. An east wind always brings these Gulls up the Thames in great numbers, and the longer it lasts and the harder it blows the farther up they seem to travel. On the 19th of August, 1886, I saw a procession of twelve fly solemnly over our house at Blackheath.

Mr. Howard Saunders says few species differ so much in individual size. He says it appears to be a species which attains its greatest development in the north and east, and deteriorates in size as it ranges south and west. He gives its range, "Throughout the Palæarctic region, but very rare in Iceland; once in Labrador." On an Australian voyage we should be followed by them all down the River Thames and the English Channel.

The bird is thus described in Yarrell's 'British Birds' (vol. iii. p. 454):—"In the adult

---

* In September, 1886, the electric launch 'Volga' passed a Common Gull between Dover and Calais so sound asleep that they lifted it on board.

bird in summer the bill is greenish grey at the base, towards the point yellow; irides dark brown, edges of the eye-lids red; the whole head and neck pure white; the back and all the wing-coverts pearl-grey, secondaries and tertials the same, but broadly edged and tipped with white; primaries black on the outer web, with a small portion of pearl-grey at the base of the inner web, the proportion of grey increasing on each primary in succession; the first and second primary with a patch of white on both webs near the end, but the extreme tips of both are black; the third, fourth, fifth, have white tips, but the first set of primary quill-feathers, which the young bird carries for the first fifteen months, have no white at the tips. Few birds moult their first set of quill-feathers in their first autumn. Tail-coverts and tail-feathers pure white; chin, neck in front, breast, and all the under surface of the body and tail pure white; legs and feet dark greenish ash. The whole length of an old male, eighteen inches and a half; of the wing, from the point, fourteen inches and a half. The length of an old female about half an inch less; and of the wing, half an inch less. In the winter the whole head and the sides of the neck are streaked and spotted with dusky brown and ash-brown. A young bird in its first autumn has the basal portion of the bill yellowish brown, the part anterior to the nostrils nearly black; irides dusky; head, sides of the neck, the ear-coverts and occiput dull white, mottled with greyish-brown; the back, wing-coverts, secondaries and tertials brownish-ash; the feathers edged with paler brown; a few bluish-grey feathers on the centre and sides of the back; the primaries nearly black, both as to the shafts and greater part of the webs, all but the front being tipped with brown; upper tail-coverts dull white; tail-feathers white, the outer half black, except the outer feather on each side, which has the outer web white; chin and throat white; neck in front, the breast, and all the under surface of the body mottled with light ash-brown, on a ground of white; legs and feet pale greyish-brown; the claws black."

This is one of the Gulls that occasionally build in trees, but never high up. As a rule, they build in marsh or rock in some slight depression, which they fill up with grass or sea-weed. They lay two or three eggs of a dark olive-brown, blotched over with black. All my specimens are nearly exactly alike.

THE LESSER BLACK-BACKED GULL (*Larus fuscus*).—I had two immature specimens of this species for some time in a garden in the hope of noting their change of plumage. Unfortunately their thieving propensities more than counterbalanced my ornithological zeal, and I was obliged to get rid of them before they arrived at the adult state. They would eat anything, or try to; nothing came amiss,—bread, meat, bones, newspapers, golf balls and tennis balls; one and all were taken to a large bowl of water to be well soaked, and then, if possible, swallowed. Of fish they were inordinately fond, and always kept close at hand if anybody was fishing on the lake, in the hope of getting a roach or dace thrown to them. If we had a rat hunt they invariably joined the dogs, and carried off the dead rats to their favourite bowl of water. On the journey the larger one would start off with his dainty, puffed out with feathers on end like a porcupine, growling and muttering in a fearfully defiant manner, and followed at a respectful distance by his mate. This was all very well; but they finally took to eating young ducks, chickens, and hen's-eggs; so were summarily dismissed. Though pinioned, they could fly from the lake up to our house; about two hundred yards. This they always did on being called for their one o'clock meal.

This species is very common all round our shores, and is perhaps the most numerous of our British Sea Gulls. They are not much appreciated by the fishermen, as they assemble in vast numbers and steal the fish out of the nets in the most barefaced way. Mr. H. Saunders says they are met with in "the North of Europe, the Faroes, the Baltic, Russia as far east as Archangel, the British Isles, the French coast, and the Canaries." We might therefore frequently fall in with them on the commencement of an outward voyage to Australia.

In Yarrell's 'British Birds' (vol. iii. p. 467) the bird is thus described:—"The adult bird in summer has the bill yellow, the inferior angle on the lower mandible red; irides straw-colour; head, and the whole of the neck all round, pure white; back, wing-coverts, and all the wing-feathers dark slate-grey, the tips only of some of the longer scapulars and tertials being white, and white tips to the shorter primaries; upper tail-coverts and tail-feathers white; breast, belly, and all the under surface of the body and tail, pure white; legs and feet yellow. The whole length, twenty-three inches; from the anterior joint of the wing to the end of the longest quill-feather, sixteen inches. In winter the head and neck are streaked with dusky brown. A young male at one year old has the base of the bill pale brown, the rest horny-black; irides dark brown; head, sides and back of the neck white, streaked longitudinally with dusky-brown; back, and all the wing-coverts and the tertials ash-brown, the feathers margined with white, but the shaft of each feather deep brown, forming a dark line down the centre; primaries and secondaries blackish-brown, without any white at the tips; upper tail-coverts white, tail-feathers blackish-brown, varied with some white, the central feathers having the most dark colour, the outside ones the most white; chin and neck in front white; breast, belly, flanks, and under tail-coverts white, mottled with dusky-brown; legs and feet light brown."

Mr. Howard Saunders says:—"The distinguishing characteristics of the adult of this species are its dark slate-coloured mantle, chrome-yellow legs and feet, and the shortness of the foot compared with the tarsus."

Concerning its nesting, Mr. Hewitson says that these birds "appear to prefer those islands which are the most rocky, and upon which there is the least herbage, and, though they have their choice, very few of them deposit their eggs upon the grass, and yet they rarely lay them without making a tolerably thick nest for their reception; it is of grass, loosely bundled together in large pieces, and placed in some slight depression or hollow of the rock. They lay two or three eggs, varying in their shades of colour from a dark olive-brown to a light drab, thickly spotted with ash-grey, and two shades of brown; the length of the egg about two inches ten lines, by one inch and eleven lines in breadth."

Larus dominicanus is a sort of southern representative of the **Lesser Black-backed Gull.** Mr. Howard Saunders says it is found in New Zealand, Kerguelen Island, and the other islands between it and Cape of Good Hope; also the Falkland Islands, Patagonia, &c. He says:—"The deep *brown*-black of the mantle, as distinct from the *slate*-black of *L. fuscus*, and its strong bill and larger size will distinguish *L. dominicanus* from that species; it is smaller than *L. marinus* (Great Black-backed Gull), has a different pattern of primaries, and has *olivaceous-coloured* legs and feet."

Larus belcheri is also much like our Lesser Black-back, only stouter, and with a black band on its tail. On an Australian voyage it might be met with off Cape Horn. The immature bird has a dark hood.

Larus scoresbii might also be met with off Cape Horn. Mr. Saunders says it is a very well-marked species, from its short, stout, crimson bill, and coarse legs and feet, the web of the latter being a good deal incised. Like *L. belcheri*, it has a dark hood when immature.

GREAT BLACK-BACKED GULL (*Larus marinus*).—This large Gull is fairly common along the mouth of the Thames, where it remains all the year round. At Westgate-on-Sea, in the month of July, they were about the only Gulls I saw, and very few of them. (Perhaps this is because there are so few shell-fish on these shores.) They were all in immature plumage, and called by the local fishermen "Grey Gulls" or "Cobs." This immature appearance lasts three years, and in some places they are then called "Wagels." I shot one of these, together with an adult bird, from a bathing-machine, at Aldborough, Suffolk, both of which are now in my collection of British birds. In Mr. E. T. Booth's wonderful collection, at Brighton, there are some adult specimens, represented attacking a Highland lamb. In his Catalogue he says he has seen wounded mallard and wigeon fly from the attack of these Gulls, and pitch close to the shooters for safety. Mr. Saunders gives their range

U

as follows:—"Northern and temperate Europe and Iceland (breeding), visiting the Mediterranean in winter, as far as Greece; the Canaries, and probably the Azores."

The bird is thus described by Yarrell, 'British Birds' (vol. iii. p. 474):—"The adult bird in summer has the bill pale yellow, the inferior angle of the under mandible reddish-orange, the whole bill very large and strong; the irides straw-yellow, the edges of the eyelids orange; head and neck pure white; back, wing-coverts, scapularies, secondaries, and tertials lead-grey, the feathers of the three latter series ending in white; primaries nearly black, the first and second quill-feathers with a triangular white patch, forming the end of each feather, the second quill-feather having a black spot in the white; all the others tipped with white, the inner broad webs being lead-grey; upper tail-coverts and tail-feathers pure white; chin, throat, breast, belly, and all the under surface of the body and tail pure white; legs and feet flesh-colour. In winter the crown of the head and the occiput are slightly streaked with ash-grey. The whole length of an adult male is thirty inches, and sometimes rather more; the wing, from the carpal joint to the end of the longest quill-feather, twenty inches. The female measures twenty-seven inches, and her wings nineteen inches."

They construct a grassy nest on marsh or rock, and lay three eggs of yellowish-greenish-brown, blotched with slate-grey and dark brown; three inches two lines in length, by two inches and four lines in breadth. This is the only "Black-back" which, when adult, has *flesh-coloured* legs and feet.

The southern representative of this bird is the Pacific Gull, *Larus pacificus*, which is confined to Australia. It has a very deep and powerful bill, of an orange-colour, and yellow legs. In the adult the tail is crossed by a black band. Its flight, too, is superior to our Great Black-back, otherwise they are much alike. We should be pretty certain to meet this species on nearing Australia.

HERRING GULL (*Larus argentatus*).—Mr. E. T. Booth says that from his own observation the farmer rather than the game-preserver would have a right to complain of the damage caused by this species. He has noticed that when frightened (as, for instance, when fired at) they throw up grain. It does not, however, belie its English name, as it is extremely fond of fish; and Mr. Yarrell tells us it is called "Pescatore" by the Italian fishermen. I had a tame one for some years that spent most of the summer months on a lake, feeding on the smaller fish, and only coming to the house to be fed in the winter. It pouched all its food, and then used to walk off, looking suspiciously all around, with its neck puffed out nearly the size of its body. This pouching seems common to all the *Laridæ*, also the habit of throwing it up when frightened. This is known by the Skua Gull, and taken advantage of whenever an opportunity occurs. A raven, after pouching an enormous amount of food, hides it away in various holes; but I never saw a Gull do this.

The Herring Gull is thus described by Yarrell ('British Birds,' vol. iii. p. 470):—"The adult bird in summer has the bill yellow, the angle of the under mandible red; edges of the eyelids orange, the irides straw-yellow; head and neck all round pure white; the back, and all the wing-coverts uniform delicate French grey; tertials tipped with white; primaries mostly black, but grey on the basal portion of the inner web; the first primary with a triangular patch of white at the end, the second and third with smaller portions of white; upper tail-coverts and tail-feathers pure white; chin, throat, breast, belly, and all the under surface of the body and tail, pure white; legs and feet flesh-colour. The whole length from twenty-two inches to twenty-four and a half, depending upon age and sex; the wing from sixteen inches and a half to seventeen and a quarter. In winter the adult birds have the head streaked with dusky-grey. Young birds resemble the young of the Lesser Black-backed Gull, but the legs and feet are more livid in colour."

Mr. H. Saunders gives its range as follows:—"The North-west of Europe from the Varanger Fiord, the Baltic, the western coasts down to North Africa, the Azores (where it breeds), Madeira, and the Canaries." He also says:—"There can be no doubt that examples from northern latitudes have a somewhat lighter mantle than those from more temperate regions, although the transition is very gradual."

Mr. Yarrell says:—"These Gulls make a nest of grass on the ledges and other flat portions of the cliff near the top, where they lay three eggs, which closely resemble those of the Lesser Black-backed Gull. They are of a light olive-brown, spotted with two shades of dark brown, and measure two inches and a half in length, by one inch and three-quarters in breadth."

GLAUCOUS GULL (*Larus glaucus*).—This is really a Gull of the Arctic regions, but goes southwards in the winter, and visits the British Isles. Still it is not likely we shall fall in with it on an Australian voyage, so I will merely say it is a Great Black-backed Gull in size and shape, only all white. Yarrell calls it the Large White-winged Gull.

The ICELAND GULL (*Larus leucopterus*) is still more arctic, and is, like the Lesser Black-backed Gull, only white. Yarrell calls it the Lesser White-winged Gull. It has been taken in this country.

GOULD'S SILVER GULL (*Larus novæ-hollandiæ*).* — This bird, the Silver Gull (*Larus scopulinus*), and *Larus hartlaubi*, are all very much alike, and very closely related. Gould's Silver Gull is the largest and has less black on the three outer primaries, and should be met with off Sardinia. The Silver Gull is essentially a New Zealand Gull. It is thus described by Gould ('Birds of Australia,' vol. ii. p. 388):—"The two sexes are precisely alike in colour. Head, neck, and all the under surface and tail, white; back and wings delicate grey; primaries white, eccentrically marked with black, largely on their inner and narrowly on their outer webs, and largely tipped with the same hue, with a slight fringe of white at the extremity; eye-lash, bill, legs, and feet, deep blood-red; nails black; irides pearl-white." *Larus hartlaubi* would be met with off the Cape of Good Hope. Mr. Saunders says it may be recognised from its two close allies by its smaller size, proportionately longer and slenderer bill, which is of a rich crimson, and by the more sooty colour of the under wing-coverts, especially along the carpal joint.

BLACK-HEADED GULL (*Larus ridibundus*).—I used to watch these birds with great interest between the intervals of golf, at St. Andrew's. They seemed regularly to follow the tide out to its furthest limits, and then fly out to sea a short distance and settle in enormous flocks till the next ebb; so that we could always tell the state of the tide by their behaviour. They seemed to make good use of their time on shore, as the sand was literally covered with smashed-up shells. In summer their black heads and red bills, legs, and feet made them very conspicuous. This black head is a real change of colour, and not a moult.

The LAUGHING GULL (*Larus atricilla*) is almost identical with the Black-headed Gull, but is slightly larger. It may, however, always be recognised by its black primaries. One was obtained by Col. Montagu at Winchelsea, Sussex.

The MASKED GULL (*Larus melanocephalus*) was also once obtained in England. It much resembles a small Black-headed Gull, only the black is more like a mask than a hood.

All these three species might be met with in the English Channel; though the Laughing Gull, being an American species, would certainly be a *rara avis.*†

LITTLE GULL (*Larus minutus*).—This Gull is the smallest of all the genus, and is a rare visitant to Great Britain. Yarrell says that on more than one occasion, when shot in this country, it was associated with Terns.

---

* Mr. J. A. Froude, in 'Oceana,' says, on nearing Melbourne, "The Albatross had left us; we were attended now by flights of the small, beautifully white Australian Gull."

† A Norfolk gamekeeper once told me he had shot a Woodpecker, called the "Rara-avis."

A specimen in summer plumage is thus described in Yarrell's 'British Birds':—"Bill reddish-brown; irides very dark brown; the whole of the head and the upper part of the neck all round is black; the neck below white; the back, wing-coverts, and wings, uniform pale ash-grey; the outer primaries darker grey, with white at the end and on the inner margin of the web; upper tail-coverts and tail-feathers white; the tail in form square at the end, and all the under surface of the body and under tail-coverts white; legs, toes, and membranes vermilion. An adult bird in winter has the bill almost black; irides dark brown; forehead and upper part of neck in front, and on the sides, pure white; occiput and nape of the neck streaked with greyish black on a white ground; a dusky spot under the eye, and an elongated patch of dusky black falling downwards from the ear-coverts; all the other parts as in summer." Whole length, ten inches and one-eighth. On an Australian voyage we should only expect to see this Gull in British waters.

BONAPARTEAN GULL (*Larus philadelphia*).— Another Gull with a brown head in summer turning white in winter. Mentioned by Yarrell in 'Appendix to British Birds' as a rare winter visitant to our shores. Much like a Tern in its movements. Thus described by Yarrell:—"Neck, tail-coverts, tail, whole under plumage, and interior of the wings, pure white; hood greyish-black, extending half an inch over the nape, and as much lower on the throat; mantle pearl-grey, this colour extending to the tips of the tertiaries, secondaries, and two posterior primaries; the anterior border of the wings white; the outer web of the first primary, and the ends of the first six are deep black, most of them slightly tipped with white; the inner web of the first primary, with the outer webs of the three following ones, with their shafts, are pure white; bill shining black; inside of the mouth and the legs bright carmine-red; irides dark brown. Whole length, fourteen inches to fifteen inches and a half; wings, from the bend to the end of the longest quill-feather, ten inches. The female is a little smaller than the male." We should only expect to meet this little Gull at the very commencement of the voyage to Australia; and as this is an American species it would be very unlikely even there.

SABINE'S GULL (*Nema Sabinii*).—The distinguishing character of the genus *Nema* is the forked tail, giving the Gull a Tern-like appearance. Mr. Saunders only includes two in the genus, and of those two we might possibly meet this one in British waters, as it has been obtained in Great Britain. It has a dark-coloured head in summer, which it loses in winter. Resembles the Black-headed Gull in appearance, only has primaries nearly quite black, black and yellow bill; legs, feet, and claws black; a forked tail, and is only thirteen inches long.

x

CUNEATE-TAILED GULL (*Rhodostethia rosea*).—This arctic Gull has a genus all to itself, on account of its cuneate tail, by which it may always be recognised. It is often called Ross's Gull. Yarrell includes it in his 'Appendix to British Birds,' so it would be possible to come across it in British waters. It has a reddish white head, neck, and under parts; pearl-grey back; black bill; and red legs and feet. Whole length, fourteen inches.

IVORY GULL (*Pagophila eburnea*).—Another arctic Gull and very rare visitant to Great Britain that has a genus all to itself. Mr. Howard Saunders says:—"The short stout bill, coarse rough feet with serrated membranes, much excised webs, and strong curved claws, appear to entitle this species to generic separation."

It is thus described in Yarrell's 'British Birds,' vol. iii. p. 451:—"The adult bird in summer has the bill greenish grey at the base and about the nostrils, the anterior portion yellow; the irides brown; eyelids red at the edge; the whole of the plumage, including the wing and tail-feathers, a pure and delicate white; the legs short and black. Whole length from sixteen to eighteen inches."

KITTIWAKE GULL (*Rissa tridactyla*).—The genus *Rissa* consists of two species of Gulls, only one of which we should expect to meet on an Australian voyage. The Kittiwake may always be known by the absence of any hind toe, which is represented by a small tubercle only. On this account the species is called *tridactyla*, three-toed. The name Kittiwake is, from the peculiar note of the bird, supposed to resemble that three-syllable word.

This Gull is common all round the British coast, but hardly ever met with inland. It builds in the rocks, and lays three eggs of a light greenish-grey, spotted with ash-grey and brown. They do not, however, seem to be very particular about the material, as I was told that once, when a ship bringing home tobacco was wrecked off Scilly, all the Kittiwakes far and wide used the tobacco for their nests.

The bird is thus described by Yarrell ('British Birds,' vol. iii. p. 447):—"The adult bird in summer at the breeding-station has the bill greenish yellow, the mouth inside orange; the irides dark brown; the head and the neck all round pure white; back and wings French grey, the secondaries and tertials tipped or edged with white; the outer margin of the first primary quill-feather black, the next three tipped with black, the fifth with a black patch near the end, but the extremity white; tail-coverts and tail-feathers pure white; chin, throat, breast, and all the under surface of the body and tail pure white; legs rather short, and dusky in colour, the toes and interdigital membranes also; the hind toe only a small tubercle, without any projecting horny nail or claw. The whole length fifteen inches and a half; from the anterior joint of the wing to the end of the longest quill-feather, twelve inches.

"The adult bird in winter has the lower part of the neck behind French grey, like the back; the occiput, top of the head, and the region of the ear-coverts streaked with dark grey, the other parts as in summer. Young birds of the year have the bill black; the irides dusky, almost black; upper part of the head white; the occiput and nape with a few dusky grey patches on a white ground; the lower part of the neck behind marked by numerous blackish grey feathers, forming transverse crescentic bands; back, scapularies, great wing-coverts, and secondaries French grey; point of the wing, and the series of smaller wing-coverts nearly black, forming a conspicuous dark stripe down the wing when closed, and across it when expanded; tertials French grey, with a spot of black near the end, the inner broad web varied with white; tail-coverts and tail-feathers white, the latter black at the end, forming a conspicuous transverse bar; the middle tail-feathers having the largest portion of black, the outer tail-feather on each side the smallest; chin, neck, breast, and all the under surface of the body pure white; under tail-coverts white, tail-feathers white at the base, ending in dark or lead-grey; legs, toes, and membranes pale brown."

Mr. Howard Saunders gives its range as follows:—"Arctic region, and along the sea-coasts of the Subarctic region, down to about 40° N. lat., breeding perhaps even in the Canaries; in winter it is abundant about the Azores, Canaries, and opposite coast of Africa. In America it is found on both Atlantic and Pacific coasts, but does not seem to extend far down the latter, nor to Japan or China, even in winter." There are therefore numerous places where we might expect to fall in with it on an Australian voyage.

CHAPTER III.

THE SKUA GULLS.

" They were, in truth, great rascals, and belonged to that *genus* who find things before they are lost."—GRIMM.

HERE are six distinct species of Skuas, four of which are included in our British fauna. They are far more courageous than the true Gulls, and are all armed with a powerful hooked beak and claws. Their now generally adopted generic name is *Stercorarius;* though *Lestris*, "a robber" or "pirate-vessel" is often used, and is certainly very appropriate. They are sometimes called "Parasitic Gulls," as to a great extent they live by robbing their weaker brethren. The name of "Skua" is from the note of the bird, which sounds like "skui." All the Skuas are thoroughly oceanic in habit, and are found in the higher latitudes of both hemispheres.

THE COMMON SKUA (*Stercorarius catarrhactes*).—The illustration, Pl. VI., Fig. 10, of the head and feet of this Skua Gull is the life-size portrait of a specimen in my collection of British birds. Mr. Howard Saunders says the range of this species is the most restricted of any member of this family which breeds in the northern hemisphere. As a species he says it is nowhere abundant, and of late years its numbers in the Faroes and Shetland Islands have so seriously diminished as to render its speedy extermination there extremely probable.

Yarrell says its breeding-stations with us are probably confined to Shetland. Quoting Mr. Dunn, he mentions three, viz.:—"Foula, Rona's Hill, and the Isle of Unst. In the latter place it is by no means numerous, and is strictly preserved by the landlords on whose property it may have settled, from a supposition that it will defend their flocks from the attacks of the Eagle. That it will attack the Eagle if he approaches their nests is a fact I have witnessed, and once saw a pair completely beat off a large Eagle from their breeding-place on Rona's Hill. The flight of the Common Skua is more rapid and stronger than that of any other Gull. It is a great favourite with the fishermen, frequently accompanying their boats to the fishing-ground, which they consider a lucky omen, and in return for its attendance they give it the refuse of the fish which are caught."

Yarrell also says the female lays two and sometimes three eggs, which are olive-brown, blotched with darker brown; the length two inches nine lines, and two inches in breadth. Though this Skua is essentially a "robber," Mr. Howard Saunders says it also feeds upon flesh, and especially upon the Kittiwake Gull.

As this bird has been shot in both Cornwall and Devonshire, it might be met with in the English Channel. It is thus described by Yarrell ('British Birds,' vol. iii. p. 484):—

"In this species the bill and its cere are black; irides dark brown; the whole of the head and neck dark umber-brown, slightly varied by streaks of reddish-brown; back, wings, and tail dark brown; scapulars and tertials margined with pale reddish-brown; wing-primaries blackish-brown, rusty brownish-white at the base; the two middle tail-feathers a little longer, and rather darker in colour than the others; chin, throat, neck in front, breast, and all the under surface of the body uniform clove-brown; legs, toes, and their membranes black; the tarsi scutellated in front, reticulated behind; the inner claw the strongest and the most curved. The whole length, twenty-four to twenty-five inches; the wing from the anterior bend, sixteen inches. The female is rather smaller than the male, but otherwise the sexes do not differ much in appearance; nor does this species assume by age the lighter colours peculiar to the other species of this genus."

POMATORHINE SKUA, *Stercorarius pomatorhinus.* — This Skua is perhaps better known as the Pomarine Skua, and Seebohm, in describing it, says that whatever *pomarinus* may mean, they are by far the most *marine* of these very marine birds. In Great Britain itself it is rare, though large flocks may every season be met with a few miles off our coasts. It is common in the Arctic Regions, and Seebohm says may be found at the Cape of Good Hope. We might therefore on an Australian voyage fall in with it in the English Channel or at the Cape of Good Hope. Though subject to changes in plumage, it may always be recognised by the legs and base of the toes being yellow. Mr. Howard Saunders says the dark pectoral band evidently becomes narrower with increasing age, until it is totally lost and the bird is pure white from the chin to the abdomen. It is on rather more slender lines than the Common Skua, which it resembles in its habits, preferring a pirate's life to one of industry.

Booth, in his well-known 'Catalogue of Birds,' says the Pomarine Skua, though occasionally compelling the Kittiwake to provide it with food, more commonly attacks the larger species of Gulls.

Describing this bird Yarrell, in 'British Birds' (vol. iii.), says:—"The bird is said to form a rude nest of grass and moss, which is placed on a tuft in marshes, or on a rock, and to lay two or three eggs; these, as figured by Naumann and Buhle, are of a uniform pale green, the larger end blotched and spotted with two shades of reddish-brown; the length two inches three lines, by one inch six lines and a half in breadth. In the young bird the cere and base of the bill are greenish-brown, the curved point black; the irides very dark brown; feathers of the head and neck clove-brown, with narrow margins of wood-brown; back, scapulars, tertials, and upper tail-coverts umber-brown, each feather margined with wood-brown, these margins being broadest on the tertials, the lower part of the back, and the upper tail-coverts; great wing-coverts nearly uniform umber-brown; wing-primaries blackish-brown, the shafts of these feathers, and a considerable portion of the inner webs white; tail-feathers umber-brown, the two middle tail-feathers in this young bird not more than half an inch longer than the nest-feather on each side; chin, throat, breast, belly, and vent mottled with buff-coloured brown, produced by narrow alternate transverse lines of clove-brown and wood-brown; under tail-coverts broadly barred across with umber-brown and wood-brown; legs and base of the toes yellow, anterior part of the toes and their intervening membranes black. The whole length of this bird to the end of the tail-feathers next the central pair, twenty inches; wing from the anterior bend, fourteen inches and a quarter. The comparative measurements in an adult bird would be twenty-one inches, and fifteen inches.

"I have seen a specimen of the Pomarine Skua in the collection of Mr. Bond, which was obtained alive when a young bird in the varied plumage of its first year, which assumed the uniform chocolate-brown plumage during its second year; some specimens barred across the breast have been named *Lestris striatus*, as noticed by Mr. Eyton, and I have seen two fine old birds, dove-grey on the back, with the head black, the neck all round and the breast yellowish-white, with the central feathers elongated, showing that the Pomarine Skua is subject to all the changes of plumage which have been so frequently observed in the more common species, Richardson's Skua."

RICHARDSON'S SKUA (*Stercorarius crepidatus*). — This bird generally goes by the name of the "Arctic Skua," referring to which fact Seebohm says it is the least Arctic of all the Skuas. It is also sometimes called the "Parasitic Gull,"* concerning which Mr. Salmon says:—"It is very amusing to see this bird chasing the Kittiwake, which it compels to disgorge its food, and before this food reaches the water or land this pirate-bird catches

---

* Though one would imagine this name should be given to Buffon's Skua (*S. parasiticus*).

it. This appears to be the only means of subsistence with this *Lestris*, as we never observed them fishing like the rest of the Gulls." But neither of these names are very distinguishable, as all the European members of this subfamily are both *arctic* and *parasitic*.

Mr. Booth, in his 'Catalogue of Birds,' says:—"This bird may still be found breeding in many parts of the North of Scotland and the adjacent islands, the nest being placed on the open moor. On land, as at sea, the Arctic Skua for the most part procures its food by robbery; those that I have seen on Strathmore usually persecuting the unfortunate Common Gulls that have the misfortune to nest in the same locality. Fish, as a rule, is their diet, but this they occasionally vary with eggs, swallowing, I believe, the whole or the greater part of the shell, as I have often noticed castings composed entirely of egg-shells on the mounds where these birds are in the habit of resting."

Concerning their extraordinary variations of plumage, Mr. Booth says there is no rule for the colouring of either sex. He also gives the following account of their thievish propensities:—"In the autumn these birds are very numerous off the northern coast wherever Kittiwakes are plentiful. When the boats are hauling their long lines for 'haddies' and whiting, hundreds of Gulls are attracted to the spot for the fish that fall from the hook while being lifted on board; these they snatch up within a foot or two of the boats, but are frequently forced to disgorge, should a Skua be near at hand. The robber appears to take no notice of the Gull if sitting on the water, beyond watching it intently; but the moment it rises on wing he attacks it."

Mr. Yarrell says the eggs are two in number, olive-brown, spotted with dark brown, and the nests of dry grass and mosses in low, marshy, unfrequented heaths.

These birds have been seen in Kent, Sussex, Hampshire, Devonshire, and Cornwall; so we might expect to see them anywhere in the English Channel. It has also been met with at the Cape of Good Hope.

The following description is from Yarrell's 'British Birds' (vol. iii. p. 492):—"The young bird during its first autumn and winter has the base of the beak and the cere brownish-grey, the anterior portion conspicuously curved and black; the irides dark brown; the head and neck pale brown, streaked with dark brown; the back, wing-coverts, and tertials umber-brown, margined with wood-brown; wing-primaries brownish-black, tipped with pale brown; tail-feathers pale brown at the base, then brownish-black to the end; the central pair half an inch longer than the others; neck in front, breast, belly, and under tail-coverts pale yellowish wood-brown, mottled and transversely barred with umber-brown; legs and the base of the toes yellow, the ends of the toes and the anterior portion of the intervening membranes black, and hence called sometimes the 'Black-toed Gull'; but this is only an indication of youth; as the bird increases in age the yellow colour is lost by degrees. The next stage, which in this species, also, as in the Pomarine Skua, probably occurs in the second year, the plumage is of a uniform greyish umber-brown, the whole of the light brown margins having disappeared, and the bird has now acquired its full size, measuring from the point of the beak to the end of the long feathers of the tail twenty inches, the central pair of tail-feathers being three inches longer than the nest-feather on each side; the wing, from the anterior bend to the end of the longest quill-feather, thirteen inches and three-quarters; the middle toe and claw together the same length, or one inch and three-quarters. After this stage a few yellow hair-like streaks appear on the sides of the neck; next the sides of the neck become lighter in colour; and, advancing in age, the neck all round becomes white, tinged with yellow, the head remaining of the same colour as the back. Males and females are not distinguishable by their plumage, and as this species, like the smaller Gulls, is capable of breeding when one year old, it is observed that birds, sometimes in similar states, and sometimes in very different states as to plumage, are in pairs at the breeding-stations."

Mr. Howard Saunders says :—"It is now well known that there are two very distinct plumages to be found in birds of this species, even in the same breeding place,—an entirely sooty form, and one with light under parts,—and that white-breasted birds pair with whole-coloured birds, as well as with those of their respective varieties." He also says :— "Now the particular characteristic by which Richardson's Skua may be distinguished, at any age before that of the nestling, is that the shafts of the other primaries are conspicuously tighter than in those of Buffon's Skua, in which *only* those of the first and second primaries are white, those of the third and successive primaries being dark.

Buffon's Skua (*Stercorarius parasiticus*). — This bird, sometimes called the "Long-tailed Skua," may at once be known by the great length of its two middle tail-feathers, resembling, in fact, the Tropic-birds in this peculiarity. It is also much smaller than the rest of the genus. It is very arctic in habits, but in winter comes down south, and visits the shores of England and America; so that we might meet it anywhere along our coast.

In Yarrell's 'British Birds' (vol. iii. p. 496) is the following :—"The egg, as figured by Thienemann, is of a pale green colour, spotted with ash-grey and dark reddish-brown; the measurements are two inches in length, by one inch and five lines in breadth. In the adult bird the base of the bill, including the cere, is dark greenish-brown, the horny, curved point black; sides and back of the neck white, tinged with straw-yellow; back, tertials, wing, and tail-coverts brownish-grey; primaries and tail-feathers almost black; chin, throat, and upper parts of belly white; lower part of the belly, the vent, and under tail-coverts light brownish-grey; legs, toes, and their membranes black; the tarsi still bearing some traces of their previous yellow colour. The whole length from the point of the beak to the end of the tail-feather next the central pair, thirteen inches and a half, the central feathers extending nine inches beyond; the wing, from the anterior bend to the end of the longest quill-feather, twelve inches; the tarsus one inch and a half; the middle toe and the claw rather shorter, or one inch and three-eighths. Independently of the difference in measurements, adult birds of this species, compared with old ones of the species previously described, have the head always much darker in colour, while the back is lighter."

These four species represent our British Skuas. A fifth European species is mentioned by Yarrell under the name of *Stercorarius cepphus;* but Mr. Howard Saunders gives it as another name for the present species.

THE CHILIAN SKUA (*Stercorarius chilensis*). — There are three species of Great Skuas, *viz.*, *S. catarrhactes*, *S. antarcticus*, and *S. chilensis*. Mr. Howard Saunders says :—" The affinities of the well-defined form are decidedly with *S. catarrhactes*, and not with *S. antarcticus;* it is, indeed, a somewhat slighter bird than the former, and remarkable for its rich cinnamon-coloured under parts, wing-coverts, and axillaries." Mr. Howard Saunders tells us the cool current known as "Humboldt's current" probably brings the bird up the shores of the South Pacific from the Magellan Straits to Peru.*

ANTARCTIC SKUA (*Stercorarius antarcticus*).—This, the giant of the Skuas, is the Southern Seas representative of our own Common Skua, which it almost exactly resembles, but on a larger scale. In mid-ocean this bird and the "Spectacled Petrel" both go by the name of "Cape Hen." It is also called "Cape Hawk."

This great Skua reigns supreme over all other ocean birds, robbing right and left,— here, there, and everywhere,—just as it suits his royal fancy. I have seen them pursue nearly every member of the Petrel family, from the Great Albatross down to the Cape Pigeon, and desist only on being handed over the spoil that tempted their cupidity. They stand out in striking contrast to most other bird-followers of a ship, in habits, appearance, and movements; for while the Petrels, the Terns, and the Gulls, are sailing easily and

* This may account for the presence of the Albatross (*A. irrorata*) at Callao.

Z

gracefully round and about the vessel, this robber goes flapping hurriedly and silently along, straight from his rocky home to the ship, to return as quickly when his curiosity, hunger, or piratical ambition is satisfied. No sweeping majestically over the ocean in gigantic circles extending from either horizon; no wheeling gracefully over the broken waters astern; no games; no skylarking; nothing but an excursion, perhaps for food, but more probably from a curiosity to see what it all means, at the same time hoping it may afford a good opening for a row.

The Falkland Islands are a sure find for a "Cape Hen" (where their local name is "Cape-Egmont Hen"); so on board a sailing ship on the homeward voyage from Australia (round the Horn) those passengers interested in such matters should keep a good look-out. In my last Australian voyage I was particularly anxious to get one of these specimens, as they are smaller and somewhat different to the more northerly birds. I had often seen one astern, but only in such weather that prevented a boat being lowered to obtain a shot. Once or twice I thought I had fairly hooked one on my Albatross-gear, but on hauling in he seemed to let go his hold. On nearing the Falklands I knew my chance was at hand, so gun and line were duly prepared, and the Captain kindly arranged I should be informed immediately a Skua was in sight. Sure enough, on nearing these inhospitable shores, one came flapping out to the ship at a great pace. I happened to be Albatross-fishing at the time, and remarked how all the other birds respectfully withdrew till his Royal Highness settled alongside my bait. This he promptly tore in pieces, but altogether declined to be caught; and the same game was enacted over and over again. It was like fishing for big-pond Carp, when they will go on sucking off your most carefully-prepared baits. After having eaten all he wanted he flapped off home again, as, though my gun was ready prepared for him on the after-skylight, it would have been useless firing. However, the next day, at tiffin, the officer on watch called down the skylight, "Skua overhead!" and rushing up on deck I found our friend high in the air in hot pursuit after a Whale-bird. After waiting some little time both pursued and pursuer flew right over the ship, and though I managed to hit him right enough, he unfortunately fell on the poop-rail, and even balanced there, but not long enough to give the midshipman on watch time to grab him before he slipped off into the sea. The heavy sea on prevented our being able to lower a boat; so I lost my Skua after all, and saw no more that voyage. I have never myself seen more than a solitary specimen at a time, but I have been told they occasionally appear in pairs.

Gould, in 'Birds of Australia,' says:—"I may mention that all the specimens from the Southern Hemisphere are rather darker in colour, and somewhat larger in size, than those from the Northern. I observed no difference in the colouring of the sexes, which may be thus described:—All the upper surface blackish-brown; the feathers of the back with whitish shafts and tips; all the under surface chocolate-brown; base and shafts of the primaries white." Mr. Howard Saunders says:—" From Campbell's Island, in 54° S., 168° E., up to Norfolk Island, the Crozets, and up to the Cape of Good Hope, where Layard observed it in April, the specimens all agree in their remarkable uniformity of sooty-brown plumage, there being few, if any, striations even upon the feathers of the neck; whilst the size of some of the examples is enormous, the primaries measuring sixteen and seventeen inches from carpal joint to tips of primaries."

WITH this I close my Ocean-Bird notes, trusting they may prove of some small use to passengers, and especially to those on the great Australian route. This voyage, whether undertaken on one of the magnificent modern steamers, or on the slower but perhaps more peaceful sailing-ship, is of considerable duration; and it becomes a matter of some importance how to spend the time to the best advantage. I have therefore written this book in the hope of affording some additional pleasure to those on board, by endeavouring to interest them in the study of their numerous bird-companions.

In conclusion, I take this opportunity of tendering my warmest thanks to Mr. Osbert Salvin for so kindly allowing me to benefit by his great experience and knowledge, and also to Miss F. E. Green for her beautiful illustrations. I would also wish to acknowledge the very full use I have made of Mr. Howard Saunders' exhaustive articles on the Terns, Gulls, and Skuas.

ADDENDA.

By an oversight, the most conspicuous bird in the region of Madeira is omitted—*viz.*, the Yellow-billed or Mediterranean Herring Gull, *Larus cachinnans*.

PAGE 3. I believe *arestruz* is the recognised Portuguese for Ostrich.

PAGE 4. Figs. 5, 6, 7 are, of course, on Plate III. Figs. 6 and 7 are transposed on this Plate.

PAGE 5, 11th line from top, for "It measured *six* feet across" read "*eleven* feet."

PAGE 7. I find Moseley is not referring to Tristan da Cunha, but to Nightingale Island.

PAGE 11, 8th line from top, for "shore-loving" read "shore-going."

PAGE 11. *D. nigripes.*—Coasting from Foochow to Shanghae (near the Fisherman's Group) I once came across these black varieties.

PAGE 14. Using the word "class" here and in page 36, is of course a misprint. Birds (*Aves*) are a class. The Albatrosses are a subfamily of the great family *Procellariidæ.*

PAGE 16. SOOTY ALBATROSS.—"The birds that fly most about in these seas (the South Atlantic) are *Alcatraci*, a sort of Sea Gulls as big as Geese, of a brownish colour, with long beaks wherewith they take fish; and which they feed on, either upon the surface of the water or while they are up in the air. At night, when they are disposed to sleep, they dash themselves aloft as high as possible, and putting their head under one wing, support themselves for some time with the other; but because the weight of their bodys must needs force them down again at last, they no sooner come to the water but they retake their flight, and both which being often repeated, they may in a manner be said to sleep waking. Oftentimes it happens that they fall into the ships as they sail, and into ours there fell two one night and one another. Those who know the nature of them farther say that in time of year they always go on shore to build their nests and that in the highest places whereby they facilitate their flight, having but short feet, and those large like unto a Goose. Of this we made an experiment upon them that fell into our ship, and found that being left at liberty upon the plain deck they could by no means raise themselves."—(Father Jerom Merolla (1682), 'Voyages and Travels,' vol. i., p. 665).

PAGE 17. STORM PETREL.—'The Field' has it that "Mother Carey" is a corruption of "Madre Cara." "The Spanish sailor, invoking by this name the aid of the Virgin during a storm, fondly sees an answer to his prayer in the appearance of the Petrels as the storm abates; not knowing that the disturbance of the water at such a time brings to the surface much of the matter on which those birds feed."

PAGE 22. WILSON'S STORM PETREL.—In describing the under tail-coverts of my specimens, I mean quite white-*tipped*. The *central* under tail-coverts are black, the outer ones white on the outer web. Length to end of tail, 8 inches; wings extending 1½ inch beyond.

PAGE 25. GREAT PETREL.—For "I caught the specimen from which Fig. 5," &c., read "Plate V." In the 'Graphic' of November 6th, 1886, there is an illustration and account of H.M.S. 'Leander' running into a sleeping Sunfish off Yokohama. The account finishes thus:—"Attempts were made to get it on board, it being deftly slung, but it dropped in two, as well it might, and oilily floated astern, a Japanese fishing-boat and a Giant Petrel waiting their turn." This is very new water for a Giant Petrel. I think it must have been a Short-tailed Albatross (*Diomedea brachyura*).

PAGE 29. FULMAR PETREL.—Considered the most numerous bird in the world, and yet only lays one egg. Darwin makes a strong point of this in his remarks on the great struggle for existence. So many die off in this competition that it becomes a matter of no moment whether there are six eggs to be hatched off or only one. No doubt the Fulmar Petrel is so eminently fitted for its position in life that the solitary "chick" generally survives. This is one of the three species of Ocean Birds so common on an Atlantic voyage, Wilson's Petrel and the Great Shearwater being the other two.

PAGE 32, 7th line, *Prion turtur*, not *inetur.*

PAGE 33. In 'Origin of Species' (p. 184), Darwin says:—"In the genus *Prion*, a member of the distinct family of the Petrels, the upper mandible is furnished with lamellæ, which are well developed and project beneath the margin; so that the beak of this bird resembles in this respect the mouth of a whale." This very clearly explains their being called "Whale-birds" at sea, though I have never met a sailor who knew the reason.

PAGE 34. *Prion brevirostris*, Gould, Proc. Zool. Soc., 1885, p. 88.—This bird appears to frequent the Island of Madeira and neighbouring rocky islets called the Desertas. Gould says it is the only *Prion* found north of the Equator. He thus describes it:—"Upper surface delicate blue; edge of the shoulder, the scapularies, outer margins of the external primaries, and the tips of the middle tail-feathers black; lores, sides of the head, and all the under surface white, stained with blue on the flanks and under tail-coverts; bill light blue, deepening into black on the sides of the nostrils and at the tip, and with a black line along the side of the under mandible; feet light blue, the interdigital membrane flesh-colour. Total length, 10½ in.; bill, ⅝ in.; wing, 6⅜ in.; tail, 3½ in.; tarsi, 1¼ in."

PAGE 34, 9th line from bottom (SNOWY PETREL), for "Arctic bird" read "Antarctic bird."

PAGE 35. GREATER SHEARWATER (*Puffinus major*).—This bird was omitted in Part I., so I introduce it here. It is a rare visitant to Great Britain, very common on the banks of Newfoundland, and occasionally met with off the Cape of Good Hope. It almost exactly resembles *P. anglorum*, except that it is larger, being 18 inches instead of 14 inches long.

PAGE 35. *Puffinus assimilis*, Gould, Proc. Zool. Soc., 1837, p. 156.—Long. tot. unc. 11; rostri, 3¾; alæ, 6½; caudæ, 3; tarsi, 1¼. *Hab.* in Novâ Cambriâ Australi. *Obs.*—Very closely allied to *P. obscurus*, but considerably smaller.

PAGE 36. *Pelecanoides urinatrix.*—In his 'Origin of Species' (p. 142), Darwin says:—"Petrels are the most aërial and oceanic of birds, but in the quiet sounds of Tierra del Fuego, the *Puffinuria berardi*, in its general habits, in its astonishing power of diving, in its manner of swimming, and of flying when made to take flight, would be mistaken by any one for an Auk or a Grebe; nevertheless it is essentially a Petrel, but with many parts of its organization profoundly modified in relation to its new habits of life."

PAGE 37. DUSKY PETREL (*P. obscurus*).—Yarrell says, "*P. obscurus* measures 11 inches in its whole length, and 6¾ from the bend of the wing to the end of the longest quill-feather." It is found in the Mediterranean, and very common off the Cape of Good Hope. Gould says it is the representative of *P. assimilis* of Australia, which it so exactly resembles that many authors consider it the same species.

PAGE 41. GREAT FRIGATE-BIRD.—Diego Garcia consists of three islands, two of which are leased by that enterprising Australian line, the Orient Steam Navigation Company. In examining a small collection of birds from these islands, Mr. Howard Saunders, in the 'Proceedings of the Zoological

Society' for 1886, thus describes this bird:—"Common, and may often be seen chasing Terns and Boobies till they make them disgorge their fish, as described by Mr. H. O. Forbes in his recent work, I have never seen Frigate-birds fishing for themselves; they are said to do so sometimes, but very rarely. Their flight is magnificent: I have seen one wheeling round and round in circles for at least five minutes without once flapping its wings, during which time it must have covered a mile of ground." Mr. Bourne, who collected the specimens, says:—"Gannets and Frigate-birds breed at the southern end of the island; and, although they are well known to be enemies on the wing, the Frigate-bird pursuing the Gannets and compelling them to disgorge the fish they have caught, yet they nest close together without molesting one another."

PAGE 42. FRIGATE-BIRD.—After writing my own ideas of the Frigate-bird, I was much pleased to find that Darwin had commented on the fact of this web-footed bird now so rarely taking to the water. He says no one except Audubon has seen the Frigate-bird alight on the water. He goes on to show by the present formation of the web-feet that it is gradually changing into a land species, or rather that its comparatively newly acquired mode of life is working this change in its structure.

PAGE 49. GANNET.—In October, 1886, several letters appeared in the 'Field,' mentioning instances of the Gannet (*Sula bassana*) being met with in England very far inland. One that measured six feet two inches from tip to tip, and three feet from tip of beak to tip of tail, was discovered in Cumberland, forty miles from the nearest coast line.

PAGE 57.—In my Chapter on the Terns I say that it is not likely we should meet with the genus *Gygis*. They do not go farther south of the eastern shores of Australia than Moreton Bay. But travelling by an Orient steamer, and passing Diego Garcia, we should probably see the White Tern (*Gygis candida*), where Mr. Howard Saunders says it is very common, perching in the cocoa-nut trees, and laying a single egg in the axils of the leaves. Mr. Darwin, describing Keeling Island, says of this bird:—"But there is one charming bird, a small and snow-white Tern, which smoothly hovers at the distance of an arm's length from your head, its large black eyes scanning with quiet curiosity your expression. Little imagination is required to fancy that so light and delicate a body must be tenanted by some wandering fairy spirit."

PAGE 66. THE GULL-BILLED TERN (*Sterna anglica*).—Mr. Howard Saunders says the American *S. aranea*, the Australian *S. macrotarsa*, and the European *S. anglica* are all the same bird.

PAGE 80. KITTIWAKE GULL (*Rissa tridactyla*).—A well-known dealer in Leadenhall Market once showed me a Gull that completely puzzled him. It turned out to be merely an immature Kittiwake (called by some "Tarrock"), with the black beak and black markings. I believe for a long time these one-year-old birds were considered a distinct species. Of course they *ought* always to be distinguished by the *three* toes (tridactyla).

PAGE 87, foot-note, for "*A. irrorata*" read "*D. irrorata*."

INDEX.

Authors differ in their ornithological names, but I trust the following will meet with approval.

The boardship names are given within brackets. Though most of these names hold good on board all ships, I have noticed that some ships' companies have their own peculiar *nommes de mer* for certain birds.

The habitat of each species is also given in this Index, but I do not mean to infer that they are not occasionally found elsewhere.

APPENDIX.

HINTS ON SKINNING.

Dime con quién andas, decirte he quién eres: "Tell me thy friends, and I will tell thee thy character," says the old Spanish proverb. Show me your skins, an ornithologist would say, and I will do likewise. And, indeed, though almost anyone can, in some way or other, manage to despoil a bird of its plumage, it requires no common combination of qualities to perform that operation with absolute success. Neatness, manual dexterity, adaptability to circumstances, and, above all, patience, are all essential in becoming a good taxidermist. In comfortable quarters, perhaps, the difficulties are but few, but the naturalist is by no means always certain of obtaining them; and those alone who have fought against the heat and insect-pests of the tropics, or attacked their subject with fur-gloves on, in a tent with the thermometer far below freezing-point, can have any idea of the indomitable perseverance that is often necessary in the practice of their art.

In these lines, however, it is not intended to discuss the moral qualities of the taxidermist, but rather to supply him with a few hints which he may advantageously combine with them, and enable him to cope with some of the difficulties to which allusion has just been made. And, first, it is essential to be provided with proper tools. The cases usually sold for skinning purposes are of little use. A student's dissecting-case is the best foundation, and it may be obtained at any surgical-instrument maker's. It should contain half a dozen scalpels of varying sizes and *good steel*, a pair of forceps, a set of chain-hooks, and a pair of stout scissors. To these should be added two or three common skewers, from six inches to a foot in length, a pair of small tinman's shears, a large and tolerably strong scalpel for rough work, a second pair of forceps, a pair of stout nail-scissors, and various brushes for dressing the skins with arsenical soap. For birds of medium size none are so good as the brushes usually sold with the sixpenny bottles of gum at stationers' shops. Larger specimens, such as Albatrosses, will require house-painters' brushes of small size, while, for the smaller species, camel's-hair brushes should be used. The skinner should provide himself with two or three of each size. Needles of various sizes (some of them may with advantage be the ordinary curved surgical needles), tailors'-thread, tow, wadding, a tin of arsenical soap, and one of plaster of Paris, will complete his outfit; but, in places where he will be unable to obtain the loan of a fine meat-saw, a small surgical saw (such as is used in the amputation of fingers) should also be taken, if it is desired to avoid considerable trouble in the division of the wing-bones of large birds.

The above will be the necessaries required by the skinner, if his work lies within the limits of civilisation, and bulk is a matter of no consideration. Without the proper appliances he will never be able to produce good work. But in the case of travel in unexplored countries, where every ounce of baggage has to be carefully considered, it is another matter, and experience will be his chief guide. A great deal may no doubt be done with arsenical soap and a sharp pocket-knife only, and in many countries soft dry moss or lichen is so abundant as to obviate the necessity of carrying stuffing of any kind. But, if possible, the tyro should learn his work in a civilised country, and defer adventuring himself to the "shifts and expedients of camp life" until he is a past master in the art.

A word or two is necessary with regard to arsenical soap. As a rule it can be procured at almost any birdstuffer's, but it is by no means always of good quality. The following is the best recipe for its preparation :—

White Soap	1½ lb.
Arsenic	1 lb.
Salts of Tartar	8 oz.
Camphor	4 oz.
Powdered Chalk	2 oz.

Shred the soap very fine into a pot, and add as little water as is necessary to dissolve it, stirring gently over a slow fire. When well dissolved add the chalk and salts of tartar, and mix thoroughly. Take it off the fire, and add the arsenic slowly, stirring meanwhile. Pound the camphor in a mortar with spirits of wine, add, and mix the whole thoroughly. It should then be poured into small tins, and left to set. In using, it should be worked up into a good lather before applying it to the skin.

Most birds, if not quite dead on being picked up, can be killed by compression of the thorax laterally ; the thumb and fingers being placed on opposite sides of the breast immediately beneath the wings. But this plan will not answer for very large birds, which are often very tenacious of life. The best method of procedure in these cases is to sever the spinal cord immediately behind the brain with a sharp penknife, but it is an operation requiring a certain amount of anatomical knowledge and skill, in the absence of which pressure with the knee on the breast-bone is the best substitute. The bird having been killed, the colour of the iris, legs, beak, and cere (if present) should be noted, as these parts often fade with great rapidity after death. Wool should be stuffed tightly down the throat, and the nostrils and all shot-wounds that seem likely to bleed should also be plugged with the same material. The bird should then be carefully wrapped in a handkerchief or towel, or hung *by the beak* upon the collecting-stick. On reaching home the specimens should be hung up in some cool place, and carefully guarded from all possible injury by cats, ants, flies, and other enemies of the ornithologist. Too great stress cannot be laid upon the exercise of caution in this respect in tropical climates, for, in countries where the white ant is abundant, an hour or two is quite sufficient for the conversion of one's most valuable specimens into skeletons.

We will presume these dangers to have been successfully avoided, and the tyro to be seated at his table anxious to commence on his first subject. His first proceeding will be to measure its length from the beak to the tip of the tail, it being no longer stiff from *rigor mortis*, which has hitherto rendered that operation difficult. The nostrils and throat should then be re-plugged with cotton-wool, and the ears and vent likewise, a precaution which, if neglected, very often results in the complete ruin of the skin. Many specimens will have their feathers more or less soiled with blood, which is especially noticeable in the Gulls and Albatrosses, and other sea-birds in which the plumage is for the most part white. These should be cleaned with a sponge and hot water, the sponge being constantly squeezed out, and only used in the direction of the feathers. The operator must not be disheartened by the bedrabbled appearance thus produced, although he should use his sponge as dry as he can, but should persevere until the plumage becomes fairly clean. He then takes a couple of the wire skewers between the thumb and finger of his right hand, and scatters plenty of plaster of Paris on the spot from time to time with his left, beating meanwhile most energetically with the skewers. This should be no leisurely operation, or the tyro will find the feathers set together in a solid block ; but if properly carried out, and the beating kept up without intermission and as rapidly as possible for ten minutes or so, the plumage, on being shaken free from the plaster of Paris, will be found as feathery and snow-white as if it had never previously been soiled. The skinner should now carry out the following directions as closely as he can :—

Lay the bird on the table with its beak to your left. Feel for the upper end of the breast-bone, and make a parting in the feathers from that point down the middle line of the body to the vent. This is the line of your incision, and, though it does not much matter how deep you make it on the breast, you should take the greatest care to make it only skin-deep on the abdomen. Your knife cannot be too sharp. Next pinch up the edge of the skin nearest you with the forceps, and after a stroke or two with the knife you will be able to get hold of it between your thumb and fingers. Over the breast and abdomen it is usually easily separable from the body, and the handle of the scalpel or even the finger should supplement the knife if possible. Skin back as far as the wing and inside of the thigh—as far, indeed, as you conveniently can. Have a packet of shaving-papers or any thin rough paper at hand, press them on various-sized pieces. Cover the body and inside of the skin with these, and press them on ; they will adhere tightly, and prevent the edges of the feathers from becoming blood-stained. Next grasp the right leg from the outside, and push it inwards towards the body between the first two fingers of the left hand, which should be used meanwhile to retract the skin as far as possible. Cut off the leg at the knee-joint, and clean the "drum-stick." Brush it well with arsenical soap, wrap it in wool, brush the latter again with soap, and return it. Next work off the skin from the outer and back part of the thigh, using your fingers and the handle of the scalpel as much as possible. The skin ought by this time to be separated from the body on that side, nearly as far as the middle line of the back. Turn your bird round with its head to your right hand, and go through exactly the same process on the other side. You should be able nearly, if not quite, to join your former work over the back. Both legs are now finished, but the skin is adherent at the vent and tail, which you have as yet left untouched. Take a piece of twine or thin whipcord about eighteen inches in length, and tie the ends together. Make a catch-loop, and hitch it round the flesh of the thigh just below the head of the bone. Do the same on the other side, and you will thus have the two thighs connected by a loop of string by which the bird can be suspended. Have a string fastened to a hook or some other contrivance in the ceiling just above your head, of sufficient length to reach your table. By passing this through the loop above mentioned, and making fast with an adjustable knot, you will have your bird hung up at a convenient height, just level with your face. As your skinning progresses, you will of course have to alter the height from time to time by shortening up your string. The advantages of this contrivance are enormous ; but in the open air, and on other occasions where it cannot be made use of, the chain-hooks will have to be substituted, two hooks being inserted in the body and the other in your table. By this means you will be able to get a pull on the skin, much as if a second person were holding the body of the bird ; but it is by no means so convenient as the other method.

Having hung up your bird, then, proceed to skin the tail, an operation which will require all your care. Cut the gut across just before the vent, and as you retract the skin, keep your left thumb outside and underneath, gently pressing the tail upwards. A careful examination of the blunt mass before you—the "pope's nose" in a fowl—will show it to be composed of two rounded lumps on either side, with a wedge of bone between them. These lumps are the insertions of the rectrices or large feathers of the tail, and must on no account be cut, or the feathers will fall out. The bone between them has, however, to be cut across, and this is best done with the tips of the nail-scissors. A touch or two of the knife will now separate the parts, and the skin is then free up to the wings. Shorten the string, and cover the newly-exposed surfaces with paper as before. Many people break the bones of the wings before beginning to skin, as they are apt to get in one's way. It is, however, a bad plan, and practice will teach you how to overcome the slight difficulties they cause. Skinning the wing is not so simple as the leg. It should be worked first from the front and then from the back (an easy matter if the bird is suspended), until the elbow-joint is visible. Cut off the meat cleanly from the bone, and cut through the latter in its middle part. Scissors, tinman's shears, or fine saw will be required, according to the size of the bird : the latter instrument has the advantage of not splintering the bone. Brush with the soap, and wrap in wool in the same manner as the leg. Having finished the second wing, which is much easier than the first, the neck will be found to offer no difficulties, but care should be taken not to prick the jugular veins. It often requires considerable patience to get the skin over the head. In some birds, especially in the case of Ducks, Woodpeckers, and Parrots, it is an impossibility, and another method has to be adopted, which will be referred to presently. On the head you will revert to your forceps. The ear is soon reached, and it should be cut across as near the bone as you can—within the meatus itself, if possible. In small and thin-skinned birds you will be able to pull the skin out of the ear with the forceps. Having finished the ear on both sides, the eye next appears, and, from the conjunctiva or covering of the eye-ball being closely connected with the eyelid, it will have to be cut through. It appears as a thin, semitransparent membrane, which is easily recognisable by practice, but care is necessary, in the case of the novice, lest the thin lower eyelid should be mistaken for it and "button-holed"—a mistake that would be fatal to the appearance of your specimen on that side. Finish the other eye, and work back the skin till the appearance of the base of the beak warns you that you have gone far enough, and that the first part of your task is over.

Now take the bird down. Take out the eyes, but be careful not to rupture them. With your pair of strong scissors cut away a portion of the base of the brain, but do not make too large an opening. Join this with two incisions made boldly along both edges of the floor of the under jaw, but do not cut into the articulation of the latter with the head. This will separate the body, leaving the head attached to the skin, but the former should not be thrown away. Remove the brains, and clean the skull thoroughly of the meat attached to it. Dress it well with the soap, which should of course be made into a lather previous to its application. Note the size of the eyeballs, and fill the sockets with tightly screwed-up pellets of clean cotton-wool of the same dimensions. Stuff the skull with the same material. Clean your fingers, and, holding up the bird with the finger and thumb of the left hand over the orbits, cover the neck plentifully with soap, and then turn the skin right side out. This operation is facilitated by feeling for the point of the beak with the fingers of the right hand, but, if the neck is too long to enable you to reach it, a small ruler or blunt stick pushed up from the feather side will often assist you considerably. Be sure that the skin is well pulled over the head. Arrange the eyelids so as to show a good circular eye of wool, and lay the skin on its back before you—the beak to your left. Note the thickness of the neck in the carcase, and construct a similar one of wadding, but some three or four inches longer; it should be tightly rolled between the palms of the hands. Take a turn or two round the point of a skewer, having first moistened one end of the wool with soap to make it "bite." By this means you can thrust it up the neck from the opening in the body, and bring it out at the mouth; it should be well soaped previously. Free the skewer and withdraw it, and you will find that you will be able to lengthen or shorten the neck, as you think necessary, by pulling the beak out, or the feathers of the neck to the right.

You have still the wings left to finish. In small birds the amount of meat on them is so small that it may be left to dry without much risk, but all specimens larger than a Blackbird should have it removed in the following manner:—Open the wing, and fix it steady on the table by means of a couple of stout pins driven through the parts in the neighbourhood of the carpal and elbow-joints. Your incision should be made between these two points, along the whole length of what corresponds to the fore-arm in man, and on the inner surface of the wing. Having worked the skin back as far as you conveniently can, it may be kept out of the way by means of the chain-hooks, while you remove the flesh from between and around the two bones. Dress with the soap, and lay a strip or two of wool inside. It is only in the larger birds that you will require a stitch or two to keep the edges of the skin together. Your bird is now nearly finished. Remove the adherent pieces of paper from the inside of the body, and dress well with arsenical soap, especially at the root of the tail. Look well at the carcase, and construct a tightly-packed body of tow or cotton-wool as near its size and shape as you can. Insert it carefully, pull the skin together, and sew it up.

Cut off the piece of wool protruding from the mouth, and tie the beak up, or it will gape in an unsightly manner. Small birds should be placed head downwards in a cone of paper, and hung up. Large specimens should have the wings kept together by tying a piece of tape round the body, the head should be kept in the right position, and the bird left undisturbed until it is dry. The length of time that this takes varies with the climate. In excessively damp regions, like New Guinea, skins never dry thoroughly unless exposed to the sun or artificial heat; while in the uplands of Africa they will sometimes become quite stiff before you have finished removing them, and require to be damped again previous to the insertion of the body.

The sex of the specimen has next to be determined. Cut the abdomen open, and push aside the intestines until the back-bone is visible. Lying close against it, and on either side, two whitish glandular bodies—the testes—will be seen if the bird is a male. If a female, the ovaries will be found occupying the same position—a collection of minute globular bodies resembling a miniature bunch of grapes. In the breeding season these will often be found of considerable size. In some cases, especially if the bird is badly shot in that region, it may be impossible to distinguish the sex, and, if so, the fact should be recorded on the label which, on finishing your bird, you will invariably affix to its leg. Recollect that, without some indication of the locality and sex, every specimen is practically valueless from a naturalist's point of view. For almost everyone forming a collection, however small it may be, it is worth while to have these labels printed. The cost is very small,—a few shillings only,—and the trouble saved is considerable. The example here given is a very good model. On the one side the colour of the iris, bill,

E MUSEO	IRIS			
O	BILL.			O
J. JONES. LEGS		LENGTH	No.	

and legs will be noted; on the other there is a space for the insertion of the length of the specimen, and another for the number referring to any entry you may have made in your note-book. Above these are two blank lines; at the end of the first you will write the symbols ♂ or ♀, according as the specimen is a male or female, leaving the rest to be filled in afterwards with the scientific name. On the second you will note the exact locality (or the latitude and longitude, if at sea) where the bird was shot, together with the date.

Allusion has been made to the impossibility, in some cases, of getting the skin of the neck to pass over the head. To obviate this difficulty the neck will have to be severed, and the skin turned right side outwards. A longitudinal incision will then have to be made in the skin of the upper part of the neck, of sufficient size to admit of the passage of the head. The back of the neck is the best place in which to make it. The head being cleaned as in the directions already given, it should be returned, and the incision carefully sewn up with fine stitches before the cotton-wool neck is inserted.

Your specimen is now skinned, dried, and labelled. It only remains for you to store it in some place of safety, where neither cats nor insects can get at it. In Europe you will not have much difficulty in this, but in many parts of the tropics the preservation of one's skins, owing to white ants and other pests, is a constant source of anxiety. On the whole the best method is to place them, with plenty of loose camphor, in camphor-wood boxes, made air-tight by pasting fine cotton-wool on the edges. As they accumulate they can, if perfectly dry, be soldered up in tin boxes. Then, and then only, is the collector's mind at rest.

F. H. H. GUILLEMARD.

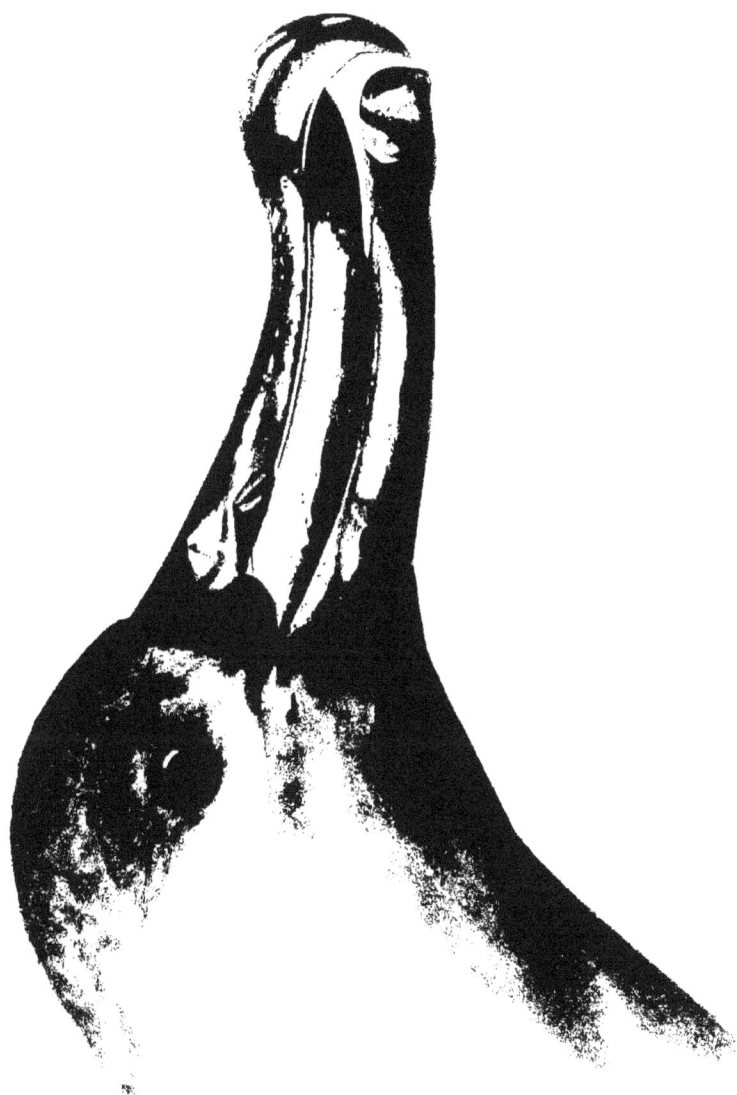

www.ingramcontent.com/pod-product-compliance
Lightning Source LLC
Chambersburg PA
CBHW021821190326
41518CB00007B/686